AF560482

Transport Geography

Transport Geography

Bimal Dhawan

RANDOM PUBLICATIONS
NEW DELHI (INDIA)

Transport Geography

ISBN 978-93-5111-416-1

Published in 2014 in India by

RANDOM PUBLICATIONS

4376-A/4B, Gali Murari Lal, Ansari Road
New Delhi-110 002
Phone : +91-11-43580356, +91-11-23289044
e-mail: randomexports@gmail.com, sales@randompublications.com, info@randompublications.com

Reprinted 2024

Type Setting by: Friends Media, Delhi-110089
Digitally Printed at : Replika Press Pvt. Ltd.

Preface

Geography and transportation intersect in terms of movement of people, goods, and information. Over time, accessibility has increased and led to a greater reliance on mobility. This trend can be traced back to the industrial revolution, although it has significantly accelerated in the second half of the twentieth-century, for various reasons. Today, societies rely on transport systems to support a wide variety of activities. These activities include commuting, supplying energy needs, distributing goods, and acquiring personal wants. Developing sufficient transport networks has been a continuous challenge to meet growing economic development, mobility needs, and ultimately to participate in the global economy. Transport and urban geography are closely intertwined, with the concept of ribbon development closely aligned to urban and transport studies. As humans increasingly seek to travel the world, the relationship transport and urban areas have often become obscured. Transportation geography measures the result of human activity between and within locations. It focuses on items such as travel time, routes undertaken, modes of transport, resource use, and sustainability of transport types on the natural environment. Other sections consider topography, safety aspects of vehicle use, and energy use within an individual's or group's journey. The purpose of transportation is to overcome space, which is shaped by both human and physical constraints, such as distance, political boundaries, time, and topographies. The specific purpose of transportation is to fulfill a demand for mobility, since it can only exist if it moves something, whether it is people or goods. Any kind of movement must consider its geographical setting, and then choose an available form of transport based on cost, availability, and space.

In terms of transport modes, the primary forms are air, rail, road, and water. Each one has its own cost associated with speed of movement, as a result of friction and place of origin and destination. For moving large amounts of goods, ships are generally used. Maritime shipping is able to carry more at a cheaper price around the world. For moving people who prefer to minimize travel time, and maximize comfort and convenience, air and road are the most common modes in usage. A railroad is often used to transport goods in areas away from water. Road transportation networks are the type of transportation that is connected with movements on constructed roads; carrying people and goods from one place to another by means of lorries, cars, etc. Water transportation is the slowest form of transportation in the movement of goods and people. Strategic chokepoints around the world have continued to play significant roles in maritime industry.

This book explores the latest advances in the field of this subject. The subject matter, both as regards the arrangement of chapters as well as contents is designed to meet the requirement of the students in several Universities.

I thank all members of my team who have helped in the preparation of the book. My special thanks go to "Random Publications" who have published the book.

—Bimal Dhawan

Contents

1

Fundamentals of Transportation/ Geography and Networks

Transportation systems have specific structure. Roads have length, width, and depth. The characteristics of roads depends on their purpose.

Modern roads are generally paved, and unpaved routes are considered trails. The pavement of roads began early in history. Approximately 2600 BCE, the Egyptians constructed a paved road out of sandstone and limestone slabs to assist with the movement of stones on rollers between the quarry and the site of construction of the pyramids.

The Romans and others used brick or stone pavers to provide a more level, and smoother surface, especially in urban areas, which allows faster travel, especially of wheeled vehicles.

The innovations of Thomas Telford and John McAdam reinvented roads in the early nineteenth century, by using less expensive smaller and broken stones, or aggregate, to maintain a smooth ride and allow for drainage. Later in the nineteenth century, application of tar (asphalt) further smoothed the ride.

In 1824, asphalt blocks were used on the Champs-Elysees in Paris. In 1872, the first asphalt street (Fifth Avenue) was paved in New York (due to Edward de Smedt), but it wasn't until bicycles became popular in the late nineteenth century that the "Good Roads Movement" took off. Bicycle travel, more so than travel by other vehicles at the time, was sensitive to rough roads. Demands for higher quality roads really took off with the widespread adoption of the automobile in the United States in the early twentieth century.

The first good roads in the twentieth century were constructed of Portland cement concrete (PCC). The material is stiffer than asphalt (or asphalt concrete) and provides a smoother ride. Concrete lasts slightly longer than asphalt between major repairs, and can carry a heavier load, but is more expensive to build and repair.

While urban streets had been paved with concrete in the US as early as 1889, the first rural concrete road was in Wayne County, Michigan, near to Detroit in 1909, and the first concrete highway in 1913 in Pine Bluff, Arkansas. By the next year over 2300 miles of concrete pavement had been layed nationally. However over the remainder of the twentieth century, the vast majority of roadways were paved with asphalt. In general only the most important roads, carrying the heaviest loads, would be built with concrete.

Roads are generally classified into a hierarchy. At the top of the hierarchy are freeways, which serve entirely a function of moving vehicles between other roads. Freeways are grade-separated and limited access, have high speeds and carry heavy flows. Below freeways are arterials. These may not be grade-separated, and while access is still generally limited, it is not limited to the same extent as freeways, particularly on older roads.

These serve both a movement and an access function. Next are collector/distributor roads. These serve more of an access function, allowing vehicles to access the network from origins and destinations, as well as connecting with smaller, local roads, that have only an access function, and are not intended for the movement of vehicles with neither a local origin nor destination. Local roads are designed to be low speed and carry relatively little traffic.

The class of the road determines which level of government administers it. The highest roads will generally be owned, operated, or at least regulated (if privately owned) by the higher level of government involved in road operations; in the United States, these roads are operated by the individual states. As one moves down the hierarchy of roads, the level of government is generally more and more local (counties may control collector/distributor roads, towns may control local streets). In some countries freeways and other roads near the top of the hierarchy are privately owned and regulated as utilities, these are generally operated as toll roads. Even publicly owned freeways are operated as toll roads under a toll authority in other countries, and some US states. Local roads are often owned by adjoining property owners and neighborhood associations.

The design of roads is specified in a number of design manual, including the AASHTO Policy on the Geometric Design of Streets and Highways (or Green Book). Relevant concerns include the alignment of the road, its horizontal and vertical curvature, its super-elevation or banking around curves, its thickness and pavement material, its cross-slope, and its width.

Freeways

A motorway or freeway (sometimes called an expressway or thruway) is a multi-lane divided road that is designed to be high-speed free flowing, access-controlled, built to high standards, with no traffic lights on the mainline. Some motorways or freeways are financed with tolls, and so may have tollbooths, either across the entrance ramp or across the mainline. However in the United States and Great Britain, most are financed with gas or other tax revenue.

Though of course there were major road networks during the Roman Empire and before, the history of motorways and freeways dates at least as early as 1907, when the first limited access automobile highway, the Bronx River Parkway began construction in Westchester County, New York (opening in 1908). In this same period, William Vanderbilt constructed the Long Island Parkway as a toll road in Queens County, New York.

The Long Island Parkway was built for racing and speeds of 60 miles per hour (96 km/hr) were accommodated. Users however had to pay a then expensive $2.00 toll (later reduced) to recover the construction costs of $2 million. These parkways were paved when most roads were not. In 1919 General John Pershing assigned Dwight Eisenhower to discover how quickly troops could be moved from Fort Meade between Baltimore and Washington to the Presidio in San Francisco by road. The answer was 62 days, for an average speed of 3.5 miles per hour (5.6 km/hr). While using segments of the Lincoln Highway, most of that road was still unpaved. In response, in 1922 Pershing drafted a plan for an 8,000 mile (13,000 km) interstate system which was ignored at the time.

The US Highway System was a set of paved and consistently numbered highways sponsored by the states, with limited federal support. First built in 1924, they succeeded some previous major highways such as the Dixie Highway, Lincoln Highway and Jefferson Highway that were multi-state and were constructed with the aid of private support. These roads however were not in general access-controlled, and soon became congested as development along the side of the road degraded highway speeds.

In parallel with the US Highway system, limited access parkways were developed in the 1920s and 1930s in several US cities. Robert Moses built a number of these parkways in and around New York City. A number of these parkways were grade separated, though they were intentionally designed with low bridges to discourage trucks and buses from using them. German Chancellor Adolf Hitler appointed a German engineer Fritz Todt Inspector General for German Roads. He managed the construction of the German Autobahns, the first limited access high-speed road network in the world.

In 1935, the first section from Frankfurt am Main to Darmstadt opened, the total system today has a length of 11,400 km. The Federal-Aid Highway Act of 1938 called on the Bureau of Public Roads to study the feasibility of a toll-financed superhighway system (three east-west and three north-south routes). Their report Toll Roads and Free Roads declared such a system would not be self-supporting, advocating instead a 43,500 km (27,000 mile) free system of interregional highways, the effect of this report was to set back the interstate program nearly twenty years in the US.

The German autobahn system proved its utility during World War II, as the German army could shift relatively quickly back and forth between two fronts. Its value in military operations was not lost on the American Generals, including Dwight Eisenhower.

On October 1, 1940, a new toll highway using the old, unutilized South Pennsylvania Railroad right-of-way and tunnels opened. It was the first of a new generation of limited access highways, generally called superhighways or freeways that transformed the American landscape. This was considered the first freeway in the US, as it, unlike the earlier parkways, was a multi-lane route as well as being limited access. The Arroyo Seco Parkway, now the Pasadena Freeway, opened December 30, 1940. Unlike the Pennsylvania Turnpike, the Arroyo Seco parkway had no toll barriers.

A new National Interregional Highway Committee was appointed in 1941, and reported in 1944 in favour of a 33,900 mile system. The system was designated in the Federal Aid Highway Act of 1933, and the routes began to be selected by 1947, yet no funding was provided at the time. The 1952 highway act only authorized a token amount for construction, increased to $175 million annually in 1956 and 1957.

The US Interstate Highway System was established in 1956 following a decade and half of discussion. Much of the network had been proposed in the 1940s, but it took time to authorize funding. In the end, a system supported by gas taxes (rather than tolls), paid for 90% by the federal government with a 10% local contribution, on a pay-as-you-go" system, was established.

The Federal Aid Highway Act of 1956 had authorized the expenditure of $27.5 billion over 13 years for the construction of a 41,000 mile interstate highway system. As early as 1958 the cost estimate for completing the system came in at $39.9 billion and the end date slipped into the 1980s. By 1991, the final cost estimate was $128.9 billion.

While the freeways were seen as positives in most parts of the US, in urban areas opposition grew quickly into a series of freeway revolts. As soon as 1959, (three years after the Interstate act), the San Francisco Board of Supervisors removed seven of ten freeways from the city's master plan, leaving the Golden Gate bridge unconnected to the freeway system. In New York, Jane Jacobs led a successful freeway revolt against the Lower Manhattan Expressway sponsored by business interests and Robert Moses among others. In Baltimore, I-70, I-83, and I-95 all remain unconnected thanks to highway revolts led by now Senator Barbara Mikulski. In Washington, I-95 was rerouted onto the Capital Beltway. The pattern repeated itself elsewhere, and many urban freeways were removed from Master Plans.

In 1936, the Trunk Roads Act ensured that Great Britain's Minister of Transport controlled about 30 major roads, of 7,100 km (4,500 miles) in length. The first Motorway in Britain, the Preston by-pass, now part of the M-6, opened in 1958. In 1959, the first stretch of the M1 opened. Today there are about 10,500 km (6300 miles) of trunk roads and motorways in England.

Australia has 790 km of motorways, though a much larger network of roads. However the motorway network is not truly national in scope (in contrast with Germany, the United States, Britain, and France), rather it is a series of local networks in and around metropolitan areas, with many intercity connection being on undivided and non-grade separated highways. Outside the Anglo-Saxon world, tolls were more widely used. In Japan, when the Meishin Expressway opened in 1963, the roads in Japan were in far worse shape than Europe or North American prior to this. Today there are over 6,100 km of expressways (3,800 miles), many of which are private toll roads.

France has about 10,300 km of expressways (6,200 miles) of motorways, many of which are toll roads.

The French motorway system developed through a series of franchise agreements with private operators, many of which were later nationalized. Beginning in the late 1980s with the wind-down of the US interstate system (regarded as complete in 1990), as well as intercity motorway programs in other countries, new sources of financing needed to be developed. New (generally suburban) toll roads were developed in several metropolitan areas.

An exception to the dearth of urban freeways is the case of the Big Dig in Boston, which relocates the Central Artery from an elevated highway to a subterranean one, largely on the same right-of-way, while keeping the elevated highway operating. This project is estimated to be completed for some $14 billion; which is half the estimate of the original complete US Interstate Highway System.

As mature systems in the developed countries, improvements in today's freeways are not so much widening segments or constructing new facilities, but better managing the road space that exists. That improved management, takes a variety of forms. For instance, Japan has advanced its highways with application of Intelligent Transportation Systems, in particular traveller information systems, both in and out of vehicles, as well as traffic control systems. The US and Great Britain also have traffic management centres in most major cities that assess traffic conditions on motorways, deploy emergency vehicles, and control systems like ramp meters and variable message signs. These systems are beneficial, but cannot be seen as revolutionizing freeway travel. Speculation about future automated highway systems has taken place almost as long as highways have been around. The Futurama exhibit at the New York 1939 World's Fair posited a system for 1960. Yet this technology has been twenty years away for over sixty years, and difficulties remain.

Layers of Networks

All networks come in layers. The OSI Reference Model for the Internet is well-defined. Roads too are part of a layer of subsystems of which the pavement surface is only one part. We can think of a hierarchy of systems.

- Places
- Trip Ends
- End to End Trip

- Driver/Passenger
- Service (Vehicle & Schedule)
- Signs and Signals
- Markings
- Pavement Surface
- Structure (Earth & Pavement and Bridges)
- Alignment (Vertical and Horizontal)
- Right-Of-Way
- Space.

At the base is *space*. On space, a specific *right-of-way* is designated, which is property where the road goes. Originally right-of-way simply meant legal permission for travellers to cross someone's property. Prior to the construction of roads, this might simply be a well-worn dirt path.

On top of the right-of-way is the *alignment*, the specific path a transportation facility takes within the right-of-way. The path has both vertical and horizontal elements, as the road rises or falls with the topography and turns as needed.

Structures are built on the alignment. These include the roadbed as well as bridges or tunnels that carry the road.

Pavement surface is the gravel or asphalt or concrete surface that vehicles actually ride upon and is the top layer of the structure. That surface may have *markings* to help guide drivers to stay to the right (or left), delineate lanes, regulate which vehicles can use which lanes (bicycles-only, high occupancy vehicles, buses, trucks) and provide additional information. In addition to marking, signs and signals to the side or above the road provide additional regulatory and navigation information.

Services use roads. Buses may provide scheduled services between points with stops along the way. Coaches provide scheduled point-to-point without stops. Taxis handle irregular passenger trips.

Drivers and *passengers* use services or drive their own vehicle (producing their own transportation services) to create an *end-to-end trip*, between an origin and destination. Each origin and destination comprises a *trip end* and those trip ends are only important because of the *places* at the ends and the activity that can be engaged in. As transportation is a derived demand, if not for those activities, essentially no passenger travel would be undertaken. With modern information technologies, we may need to consider additional systems, such as Global Positioning Systems (GPS), differential GPS, beacons, transponders, and so on that may aide the steering or navigation processes. Cameras, in-pavement detectors, cell phones, and other systems monitor the use of the road and may be important in providing feedback for real-time control of signals or vehicles.

Each layer has rules of behaviour:

- some rules are physical and never violated, others are physical but probabilistic
- some are legal rules or social norms which are occasionally violated.

2

Public Transport Systems

Public transport (also public transportation, public transit, or mass transit) is a shared passenger transportation service which are available for use by the general public, as distinct from modes such as Taxicab, car pooling which are not shared by strangers without private arrangement.

Public transport modes include buses, trolleybuses, trams and trains, 'rapid transit' (metro/subways/undergrounds etc.) and, ferries. Intercity public transport is dominated by airlines, coaches, and intercity rail. High-speed rail networks are being developed in a many parts of the world. Most public transport runs to a scheduled timetable with the most frequent services running to a headway. Share taxi offers on-demand services in many parts of the world and some services will wait until the vehicle is full before it starts. Paratransit is sometimes used in areas of low-demand and for people who need a door-to-door service.

Urban public transport may be provided by one or more private transport operators or by a transit authority. Public transport services are usually funded by fares charged to each passenger. Services are normally regulated and possibly subsidized from local or national tax revenue. Fully-subsidised, zero-fare services operate in some towns and cities.

For historical and economic reasons, there are differences internationally regarding use and extent of public transport. While countries in Old World tend to have extensive and frequent systems serving their old and dense cities, most cities of the New World have more sprawl and much less comprehensive public transport.

History

Conveyances for public hire are as old as the first ferries, and the earliest public transport was water transport: on land people walked (sometimes in groups and on pilgrimages, as noted in sources such as the Bible and Canterbury Tales) or (at least in the Old World) rode an animal. Ferries are part of Greek mythology — corpses in ancient Greece were buried with a coin underneath their tongue to pay the ferryman Charon to take them to Hades.

Some historical forms of public transport are the stagecoach, travelling a fixed route from coaching inn to coaching inn, and the horse-drawn boat carrying paying passengers, which was a feature of European canals from their 17th-century origins. (The canal itself is a form of infrastructure dating back to antiquity — it was used at least for freight transportation in ancient Egypt to bypass the Aswan cataract — and the Chinese also built canals for transportation as far back as the Warring States period. Whether or not those canals were used for for-hire public transport is unknown; the Grand Canal was primarily used for shipping grain.) The omnibus, the first organized public transit system within a city, appears to have originated in Paris, France, in 1662, although the service in question failed a few months after its founder died; omnibuses are next known to have appeared in Nantes, France, in 1826. The omnibus was introduced to London in July 1829.

Mode

Airline

An airline provides scheduled service with aircraft between airports. Air travel has high up to very high speeds, but incurs large waiting times prior and after travel, and is therefore often only feasible over longer distances or in areas where lack of ground infrastructure makes other modes of transport impossible. Bush airlines work more similar to bus stops; an aircraft waits for passengers and takes off when the aircraft is full.

Bus and Coach

Bus services use buses on conventional roads to carrying numerous passengers on shorter journeys. Buses operate with low capacity (i.e. compared with trams or trains), and can operate on conventional roads, with relatively inexpensive bus stops to serve passengers. Therefore buses are commonly used in smaller cities and towns, in rural areas as well for shuttle services supplementing in large cities. Bus rapid transit is an ambiguous term used for buses operating on dedicated right-of-way, much like a light rail.

Trolleybuses are electric buses that employ overhead wires to get power for traction. Online Electric Vehicles are buses that run on a conventional battery, but are recharged frequently at certain points via underground wires. Coach services use coaches (long-distance buses) for suburb-to-CBD or longer distance transportation. The vehicle are normally equipped with more comfortable seating, a separate luggage compartment, video and possibly also a toilet. They have higher standards than city buses, but a limited stopping pattern.

Trains

Passenger rail transport is the conveyance of passengers by means of wheeled vehicles specially designed to run on railways. Trains allow high capacity on short or long distance, but require track infrastructure and stations to be built. Urban rail transit consists of trams, light rail, rapid transit, people movers, commuter rail and funiculars.

Commuter, Intercity, and High-speed Rail

Commuter rail is part of an urban area's public transport; it provides faster services to outer suburbs and neighbouring towns and villages. Trains stop at all stations, that are located to serve a smaller suburban or town centre. The stations often being combined with shuttle bus or park and ride systems at each station. Frequency may be up to several times per hour, and commuter rail systems may either be part of the national railway, or operated by local transit agencies.

Intercity rail is long-haul passenger services that connect multiple urban areas. They have few stops, and aim at high average speeds, typically only making one of a few stops per city. These services may also be international.

High-speed rail is passenger trains operating significantly faster than conventional rail—typically defined as at least 200 kilometres per hour (120 mph). The most predominant systems have been built in Europe and Japan, and compared with air travel, offer long-distance rail journeys as quick as air services, have lower prices to compete more effectively and uses electricity instead of combustion.

Trams and Light Rail

Trams are railborne vehicles that run in city streets or dedicated tracks. They have higher capacity than buses, but must follow dedicated infrastructure with rails and wires either above or below the track, limiting their flexibility.

Light rail is a modern development (and use) of the tram, with dedicated right-of-way not shared with other traffic, step-free access and increased speed. Light rail lines are, thus, essentially modernized interurbans.

Metro/Underground/Rapid Transit

A rapid transit Metro/Underground/Elevated railway operates in an urban area with high capacity and frequency, and grade separation from other traffic.

Systems are able to transport large amounts of people quickly over short distances with little land use. Variations of rapid transit include people movers, small-scale light metro and the commuter rail hybrid S-Bahn. More than 160 cities have rapid transit systems, totalling more than 8,000 km (4,971 mi) of track and 7,000 stations. Twenty-five cities have systems under construction.

Personal Rapid Transit

Personal rapid transit is an automated cab service that runs on rails or a guideway. This is an uncommon mode of transportation (excluding elevators) due to the complexity of automation. A fully implemented system might provide most of the convenience of individual automobiles with the efficiency of public transit. The crucial innovation is that the automated vehicles carry just a few passengers, turn off the guideway to pick up passengers (permitting other PRT vehicles to continue at full speed), and drop them off to the location of their choice (rather than at a stop). Conventional transit simulations show that PRT might attract many auto users in problematic medium density urban areas. A number of experimental systems are in progress. One might compare personal rapid transit to the more labour-intensive taxi or paratransit modes of transportation, or to the (by now automated) elevators common in many publicly accessible areas.

Ferry

A ferry is a boat or ship, used to carry (or *ferry*) passengers, and sometimes their vehicles, across a body of water. A foot-passenger ferry with many stops is sometimes called a water bus. Ferries form a part of the public transport systems of many waterside cities and islands, allowing direct transit between points at a capital cost much lower than bridges or tunnels, though at a lower speed. Ship connections of much larger distances (such as over long distances in water bodies like the Mediterranean Sea) may also be called ferry services.

Operation

Infrastructure

All public transport runs on infrastructure, either on roads, rail, airways or seaways; all consists of interchanges and way. The infrastructure can be shared with other modes of transport, freight and private transport, or it can be dedicated to public transport. The latter is especially true in cases where there are capacity problems for private transport. Investments in infrastructure are high, and make up a substantial part of the total costs in systems that are expanding. Once built, the infrastructure will further require operating and maintenance costs, adding to the total costs of public transport. Sometimes governments subsidize infrastructure by providing it free of charge, just like is common with roads for automobiles.

Interchanges

Interchanges are locations where passengers can switch mode. Most interchanges are predominantly for passenger to change from being pedestrians to passengers (such as a bus stop), while each system will have a few hubs that allow passengers to change between vehicles. This may be between vehicles of the same mode (like a bus interchange), or it can be between local and intercity transport (such as at a central station or airport). Other interchange facilities include car parks and bicycle parking.

Schedules

All public transport must either operate after a predefined schedule, or operate at a sufficient frequency that travellers do not need to use a schedule to correspond with the services. Operators will publish timetables, often supplemented with maps and fare schemes to help travellers coordinate their travel. Public transport route planner online, sometimes combined with pre-sold tickets, help make planning task more user-friendly. To further aid travellers, operators often run at fixed times of the hour, so passengers only need to memorize the minutes past the hour the service leaves, and can apply that to any hour of the day.

Coordination between services at intersections is important to reduce the total travel time for passengers. This can be done by coordinating shuttle services with main routes, or by creating a fixed time (for instance twice per hour) when all bus and rail routes meet at a station and exchange passengers.

Peak and Base Period

Peak:

- A.m. peak period is the period of time in the morning when additional services are provided to handle higher passenger volumes. The period begins when normal, scheduled headways are reduced and ends when headways return to normal.
- P.m. peak period is the period in the afternoon or evening when additional services are provided to handle higher passenger volumes. The period begins when normal headways are reduced and ends when headways are returned to norma

Midday period is the period of time between the end of the a.m. peak and the beginning of the p.m. peak

Peak/base ratio is the number of vehicles operated in passenger service during the peak period divided by the number operated during the base period

Financing

The main sources of financing are ticket revenue, government subsidies and advertisement. The percentage of revenue from passenger charges is known as the farebox recovery ratio. A limited amount of income may come from land development and rental income from stores and vendors, parking fees, and leasing tunnels and rights-of-way to carry fibre optic communication lines.

Fare and Ticketing

Most—but not all—public transport required the purchase of a ticket to generate revenue for the operators. Tickets may either be bought in advance, at the time of the ride, or the carrier may allow both methods. Passengers may be issued with a paper ticket, metal or plastic token, or an electronic card (smart card, contactless smart card). Sometimes a ticket has to be validated, e.g. a paper ticket that has to be stamped, or an electronic ticket that has to be checked in.

Tickets may be valid for a single (or return) trip, or valid within a certain area for a period of time. The fare is based on the travel class, either as a function of the traveled distance, or based on a zone pricing.

The tickets may have to be shown or checked automatically at the station platform or when boarding, or during the ride by a conductor. Operators may choose to control all riders, allowing sale of the ticket at the time of ride. Alternatively, a proof-of-payment system allows riders to enter the vehicles without showing the ticket, but riders may or may not be controlled by a ticket controller; if the rider fails to show proof of payment, the operator may fine the rider at the magnitude of the fare.

Multi-use tickets allow travel more than once. In addition to return tickets, this includes period cards allowing travel within a certain area (for instance month cards), or during a given number of days that can be chosen within a longer period of time (for instance eight days within a month). Passes aimed at tourists, allowing free or discounted entry at many tourist attractions, typically include zero-fare public transport within the city. Period tickets may be for a particular route (in both directions), or for a whole network. A free travel pass allowing free and unlimited travel within a system is sometimes granted to particular social sectors, for example students, elderly, children, employees (*job ticket*) and the physical or mentally disabled.

Zero-fare public transport services are funded in full by means other than collecting a fare from passengers, normally through heavy subsidy or commercial sponsorship by businesses. Several mid-size European cities and many smaller towns around the world have converted their entire bus networks to zero-fare. Local zero-fare shuttles or inner-city loops are far more common than city-wide systems. There are also zero-fare airport circulators and university transportation systems.

Subsidies

Governments, of any variety, may opt to subsidize public transport, for social, environmental or economical reasons. Key motivations are the need to provide transport to people those who cannot afford or are physically or legally incapable of using an automobile, and to reduce congestion, land use and emissions of local air pollution and greenhouse gases. Other motives may be related to promote business and economic growth, or urban renewal in formerly deprived areas of the city. Some systems are owned and operated by a government agency; other transportation services may be commercial, but receive greater benefits from the government compared to a normal company.

Subsidies may take the form of direct payments to financially unprofitable services, but also indirect subsidies are used. This may include allowing use of state-owned infrastructure without payment or for less than cost-price (may apply for railways and roads), to stimulate public transport's economic competitiveness over private transport, that normally also has free infrastructure (subsidized through such things as gas taxes). Other subsidies include tax advantages (for instance aviation fuel is typically not taxed), bailouts if companies that are likely to collapse (often applied to airlines) and reduction of competition through licensing schemes (often applied to taxis and airlines). Private transport is normally subsidized indirectly through free roads (paid for largely by gas taxes) and infrastructure, as well as incentives to build car factories and, on occasion, directly via bailouts of automakers.

Land development schemes may be initialized, where operators are given the rights to use lands near stations, depots, or tracks for property development. For instance, in Hong Kong, MTR Corporation Limited and KCR Corporation generate profits from land development to cover the partial cost of construction, but not operation, of the urban rail systems.

Some government officials believe that use of taxpayer capital to fund mass transit will ultimately save taxpayer money in other ways, and therefore, state-funded mass transit is a benefit to the taxpayer. (Such a belief has been backed up by research, although the measurement of benefits and costs is a complex and controversial issue.) A lack of mass transit results in more traffic (perhaps, although right-wing think tanks disagree), pollution, and road construction to accommodate more vehicles, all costly to taxpayers; providing mass transit will therefore alleviate these costs.

Safety and Security

Expansion of public transportation systems is often opposed (particularly in North America) by critics who see them as vehicles for violent criminals and homeless persons to expand into new areas (to which they would otherwise have to walk). According to the Transportation Research Board, "violent crime is perceived as pandemic.... Personal security affects many peoples' [sic] decisions to use public transportation." Despite the occasional highly publicized incident, the vast majority of modern public transport systems are well designed and patrolled and generally have low crime rates. Many systems are monitored by CCTV, mirrors, or patrol.

Nevertheless, some systems attract vagrants who use the stations or trains as sleeping shelters, though most operators have practices that discourage this.

Though public transit accidents attract far more publicity than car wrecks, public transport is much safer, due to far lower accident rates. Annually, public transit prevents 200,000 deaths, injuries, and accidents had equivalent trips been made by car. The National Safety Council estimates riding the bus as over 170 times safer than private car.

Environmental

A 2002 study by the Brookings Institution and the American Enterprise Institute found that public transportation in the U.S uses approximately half the fuel required by cars, SUV's and light trucks. In addition, the study noted that "private vehicles emit about 95 percent more carbon monoxide, 92 percent more volatile organic compounds and about twice as much carbon dioxide and nitrogen oxide than public vehicles for every passenger mile traveled".

A controversial 2004 study from Lancaster University concluded that a family of four in a modern car travelling from London to Edinburgh would be more efficient than travelling in a diesel-powered UK trains. The study showed that trains had failed to keep up with the advances that the automotive and aviation industries had made in improved fuel efficiency. A representative from *Modern Railways* magazine said: Studies have shown that there is a strong inverse correlation between urban population density and energy consumption per capita, and that public transport could play a key role in increasing urban population densities, and thus reduce travel distances and fossil fuel consumption.

Going Green

Public Transportation has been a key aspect of the Green initiative. The idea of going Green, which basically entails commissioning more eco-friendly systems, is essentially new. Gases emitted by automobiles have been cited as major contributors to the issues addressed in green initiatives. A study conducted in Milan, Italy in 2004 during and after a transportation strike serves to illustrate the impact that mass transportation has on the environment. Air samples were taken between January 2 and January 9, and then tested for Methane, Carbon Monoxide, non-methane Hydrocarbons (NMHCs), and other gases identified as harmful to the environment. The figure below is a computer simulation showing the results of the study "with January 2nd showing the lowest concentrations as a result of decreased activity in the city during the holiday season. January 9th showed the highest NMHC concentrations because of increased vehicular activity in the city due to a public transportation strike."

Public Transportation allows for cars to be removed from the road. This lowers gas emissions and traffic congestions. Influenced by the previous, the state of New Jersey released *Getting to Work: Reconnecting Jobs with Transit.* This initiative, as suggested by its title, attempts to relocate new jobs into areas with higher public transportation accessibility. The initiative cites the use of public transportation as being a means of reducing traffic congestion, providing an economic boost to the areas of job relocation, and most importantly, contributing to a green environment by reducing Carbon Dioxide (CO2) emissions.

CO_2 and Energy Impact

Using Public transportation can result in a reduction of an individual's carbon footprint. A single person, 20-mile round trip by car can be replaced using public transportation and result in a net CO2 emissions reduction of 4,800 lbs/year. Using public transportation saves CO2 emissions in more ways than simply travel as public transportation can help to alleviate traffic congestion as well as promote more efficient land use. When all three of these are considered, it is estimated that 37 million metric tones of CO2 will be saved annually. Another study claims that using public transit instead of private in the U.S. in 2005 would have reduced CO2 emissions by 3.9 million metric tones and that the resulting traffic congestion reduction accounts for an additional 3.0 million metric tons of CO2 saved. This is a total savings of about 6.9 million metric tones per year given the 2005 values.

In order to compare energy impact of public transportation to private transportation, the amount of energy per passenger mile must be calculated. The reason that comparing the energy expenditure per person is necessary is to normalize the data for easy comparison. Here, the units are in per 100 p-km (read as person kilometre or passenger kilometre). In terms of energy consumption, public transportation is better than individual transport in a personal vehicle. In England, bus and rail are popular methods of public transportation, especially London. Rail provides rapid movement into and out of the city of London while busing helps to provide transport within the city itself. As of 2006-2007, the total energy cost of London's trains was 15 kWh per 100 p-km, about 5 times better than a personal car. For busing in London, it was 32 kWh per 100 p-km, or about 2.5 times that of a personal car. This includes lighting, depots, inefficiencies due to capacity (i.e., the train or bus may not be operating at full capacity at all times), and other inefficiencies. Efficiencies of transport in Japan in 1999 were 68 kWh per 100 p-km for a personal car, 19 kWh per 100 p-km for a bus, 6 kWh per 100 p-km for rail, 51 kWh per 100 p-km for air, and 57 kWh per 100 p-km for sea. These numbers from either country can be used in energy comparison calculations and/or life cycle assessment calculations.

Public transportation also provides an arena to test environmentally friendly fuel alternatives, such as hydrogen-powered vehicles. Swapping out materials to create lighter public transportation vehicles with the same or better performance will increase environmental friendliness of public transportation vehicles while maintaining current standards or improving them. Informing the public about the positive environmental effects of using public transportation in addition to pointing out the potential economic benefit is an important first step towards making a difference.

Area

Urban space is a precious commodity and public transport consumes it more efficiently than a car dominant society, allowing cities to be built more compactly than if they were dependent on automobile transport. If public transport planning is at the core of urban planning, it will also force cities to be built more compactly to create efficient feeds into the stations and stops of transport. This will at the same time allow the creation of centres around the hubs, serving passengers' need for their daily commercial needs and public services. This approach significantly reduces urban sprawl.

Social

An important social role played by public transport is to ensure that all members of society are able to travel, not just those with a driving license and access to an automobile—which include groups such as the young, the old, the poor, those with medical conditions, and people banned from driving. Automobile dependency is a name given by policy makers to places where the those without access to a private vehicle do not have access to independent mobility.

Above that, public transportation opens to its users the possibility of meeting other people, as no concentration is diverted from interacting with fellow-travellers due to any steering activities. Adding to the above-said, public transport becomes a location of inter-social encounters across all boundaries of social, ethnic and other types of affiliation.

Economic

Public transport allows transport at an economy of scale not available through private transport. Through stimulating public transport it is possible to reduce the total transport cost for the public. Time costs can also be reduced as cars removed from the road through public transit options translate to less congestion and faster speeds for remaining motorists. Transit-oriented development can both improve the usefulness and efficiency of the public transit system as well as result in increased business for commercial developments.

Well-designed transit systems can have a positive effect on real estate prices. The Hong Kong metro MTR generates a profit by redeveloping land around its stations. Much public opposition to new transit construction can be based on the concern about the impact on neighbourhoods of this new economic development. Few localities have the ability to seize and reassign development rights to a private transit operator, as Hong Kong has done. Increased land desirability has resulted around stations in places such as Washington, D.C..

Investment in public transport also stimulates the economy locally, with between $4 and $9 of economic activity resulting from every dollar spent. Many businesses rely on access to a transit system, in particular in cities and countries where access to cars is less widespread, businesses which require large amounts of people going to a same place may not be able to accommodate a large number of cars (concert venues, sport stadia, airports, exhibitions centres,...), or businesses where people are not able to use a car (bars, hospitals, or industries in the tourism sector whose customers may not have their cars).

Transit systems also have an effect on derivated businesses: commercial websites have been founded, such as Hopstop.com, that give directions through mass transit systems; in some cities, such as London, products themed on the local transport system are a popular tourist souvenir.

Conversely, the existence of a transit system can lower land values, either through influence on a region's demographics and crime rate (actual or perceived) or simply through the ambient noise and other discomforts the system creates.

Regulations

Food and Drink

Longer distance public transport sometimes sell food and drink on board, and/or have a dedicated buffet car and/or dining car. However, some urban transport systems forbid the consumption of food, drink, or even chewing gum when riding on public transport. Sometimes only types of food are forbidden with more risk of making the vehicles dirty, e.g. ice creams and French fries, and sometimes potato chips.

Some systems prohibit carrying open food or beverage containers, even if the food or beverage is not being consumed during the ride.

Smoking

In the United States, Canada, most of the European Union, Australia, and New Zealand, smoking is prohibited in all or some parts of most public transportation systems due to safety and health issues. Generally smoking is not allowed on buses and trains, while rules concerning stations and waiting platforms differ from system to system. The situation in other countries varies widely.

Noise

Many mass transit systems prohibit the use of audio devices, such as radios, CD players, and MP3 players unless used with an earphone through which only the user can hear the device. Some mass transit systems have restricted the use of mobile phones. Long distance train services, such as the Amtrak system in the US, have "quiet cars" where mobile phone usage is prohibited.

Some systems prohibit passengers from engaging in conversation with the operator. Others require that passengers who engage in any conversation must keep the noise level low enough that it not be audible to other passengers. Some systems have regulations on the use of profanity. In the United States, this has been challenged as a free speech issue.

Banned Items

Certain items considered to be problematic are prohibited or regulated on many mass transit systems. These include firearms and other weapons (unless licensed to carry), explosives, flammable items, or hazardous chemicals and substances.

Many systems prohibit live animals, but allow those that are in carrying cases or other closed containers. Additionally, service animals for the blind or disabled are permitted.

Some systems prohibit items of a large size that may take up a lot of space, such as bicycles. But more systems in recent years have been permitting passengers to bring bikes.

In Sydney, it is illegal to carry spray cans or permanent markers on public transport, as they can be used to vandalise the vehicles and stations. This rule also applies to sharp instruments that could damage the train, such as screwdrivers that could be used to make "scratchitti", a form of vandalism where tags are carved into a window.

Other Regulations

Many systems have regulations against behaviour deemed to be unruly or otherwise disturbing to other passengers. In such cases, it is usually at the discretion of the operator, police officers, or other transit employees to determine what behaviours fit this description. Some systems have regulations against photography or videography of the system's vehicles, stations, or other property. Those seen holding a mobile phone in a manner consistent with photography are considered to be suspicious for breaking this rule.

Sleeping

In the era when long distance trips took several days, sleeping accommodations were an essential part of transportation. (On land, the lodging involved was often part of the infrastructure: the inn or ryokan, which didn't move, sheltered travellers. People also slept on ships at sea.) Today, most airlines, inter-city trains and coaches offer reclining seats and many provide pillows and blankets for overnight travellers. Better sleeping arrangements are commonly offered for a premium fare and include sleeping cars on overnight trains, larger private cabins on ships and airplane seats that convert into beds. Budget-conscious tourists sometimes plan their trips using overnight train or bus trips in lieu of paying for a hotel. The ability to get additional sleep on the way to work is attractive to many commuters using public transport.

Because night trains or coaches can be cheaper than motels, homeless persons often use these as overnight shelters, as with the famous Line 22 ("Hotel 22") in Silicon Valley. Specifically, a local transit route with a long overnight segment and which accepts inexpensive multi-use passes will acquire a reputation as a "moving hotel" for people with limited funds. Most transportation agencies actively discourage this. For this and other reasons passengers are often required to exit the vehicle at the end of the line; they can board again in the same or another vehicle, after some waiting. Also, even a low fare often deters the poorest individuals, including homeless people.

Impact

Economic

Transport is a key necessity for specialization—allowing production and consumption of products to occur at different locations. Transport has throughout history been a spur to expansion; better transport allows more trade and a greater spread of people. Economic growth has always been dependent on increasing the capacity and rationality of transport. But the infrastructure and operation of transport has a great impact on the land and is the largest drainer of energy, making transport sustainability a major issue.

Modern society dictates a physical distinction between home and work, forcing people to transport themselves to places of work or study, as well as to temporarily relocate for other daily activities. Passenger transport is also the essence of tourism, a major part of recreational transport. Commerce requires the transport of people to conduct business, either to allow face-to-face communication for important decisions or to move specialists from their regular place of work to sites where they are needed.

Planning

Transport planning allows for high utilization and less impact regarding new infrastructure. Using models of transport forecasting, planners are able to predict future transport patterns. On the operative level, logistics allows owners of cargo to plan transport as part of the supply chain. Transport as a field is studied through transport economics, the backbone for the creation of regulation policy by authorities. Transport engineering, a sub-discipline of civil engineering, and must take into account trip generation, trip distribution, mode choice and route assignment, while the operative level is handles through traffic engineering.

Because of the negative impacts made, transport often becomes the subject of controversy related to choice of mode, as well as increased capacity. Automotive transport can be seen as a tragedy of the commons, where the flexibility and comfort for the individual deteriorate the natural and urban environment for all. Density of development depends on mode of transport, with public transport allowing for better spacial utilization. Good land use keeps common activities close to peoples homes and places higher-density development closer to transport lines and hubs; minimize the need for transport. There are economies of agglomeration. Beyond transportation some land uses are more efficient when clustered. Transportation facilities consume land, and in cities, pavement (devoted to streets and parking) can easily exceed 20 percent of the total land use. An efficient transport system can reduce land waste.

Too much infrastructure and too much smoothing for maximum vehicle throughput means that in many cities there is too much traffic and many—if not all—of the negative impacts that come with it. It is only in recent years that traditional practices have started to be questioned in many places, and as a result of new types of analysis which bring in a much broader range of skills than those traditionally relied on—spanning such areas as environmental impact analysis, public health, sociologists as well as economists who increasingly are questioning the viability of the old mobility solutions. European cities are leading this transition.

Environment

Transport is a major use of energy, and burns most of the world's petroleum. This creates air pollution, including nitrous oxides and particulates, and is a significant contributor to global warming through emission of carbon dioxide, for which transport is the fastest-growing emission sector. By subsector, road transport is the largest contributor to global warming. Environmental regulations in developed countries have reduced the individual vehicles emission; however, this has been offset by an increase in the number of vehicles, and more use of each vehicle. Some pathways to reduce the carbon emissions of road vehicles considerably have been studied. Energy use and emissions vary largely between modes, causing environmentalists to call for a transition from air and road to rail and human-powered transport, and increase transport electrification and energy efficiency.

Other environmental impacts of transport systems include traffic congestion and automobile-oriented urban sprawl, which can consume natural habitat and agricultural lands. By reducing transportation emissions globally, it is predicted that there will be significant positive effects on Earth's air quality, acid rain, smog and climate change.

3

An Overview of Transport Geography

Transportation Geography is the branch of geography that investigates spatial interactions, let them be of people, freight and information. It can consider humans and their use of vehicles or other modes of travelling as well as how markets are serviced by flows of finished goods and raw materials. It is a branch of Economic geography. "The ideal transport mode would be instantaneous, free, have an unlimited capacity and always be available. It would render space obsolete. This is obviously not the case. Space is a constraint for the construction of transport networks. Transportation appears to be an economic activity different from others. It trades space with time and thus money" (translated from [Merlin, 1992]).

Geography and transportation intersect in terms of the movement of peoples, goods, and information. Over time, accessibility has increased and this has led to a greater reliance on mobility. This trend could be traced back to the industrial revolution although it has significantly accelerated in the second half of the 20th century for various reasons. Today, societies rely on transport systems to support a wide variety of activities. These activities include commuting, supplying energy needs, distributing goods, and acquiring personal wants. The development of sufficient transport networks has been a continuous challenge to meet growing economic development, mobility needs, and ultimately to participate in the global economy.

Transport and urban geography are closely intertwined, with the concept of ribbon development being closely aligned to urban and transport studies. As humans increasingly seek to travel the world, the relationship transport and urban areas have often become obscured.

Transportation geography measures the result of human activity between and within locations. It focuses on items such as travel time, routes undertaken, modes of transport, resource use and sustainability of transport types on the natural environment. Other sections consider topography, safety aspects of vehicle use and energy use within an individual's or group's journey.

The purpose of transportation is to overcome space which is shaped by both human and physical constraints such as distance, political boundaries, time and topographies. The specific purpose of transportation is to fulfil a demand for mobility, since it can only exist if it moves something, be it people or goods. Any kind of movement must consider its geographical setting, and then choose an available form of transport based on cost, availability, and space.

Transportation Modes

In terms of transport modes, the primary forms are air, rail, road, and water. Each one has its own cost associated with; speed of movement as a result of friction, and the place of origin and destination. For moving large amounts of goods, ships are generally utilized. Maritime shipping is able to carry more at a cheaper price around the world. For moving people who prefer to minimize travel time, and maximize comfort and convenience, air and road are the most common modes in usage. Rail road is often utilized to transport goods in areas away from water. Also water transportation is based upon early constructioning from railroad.

" Transportation modes are an essential component of transport systems since they are the means by which mobility is supported. Geographers consider a wide range of modes that may be grouped into three broad categories based on the medium they exploit: land, water and air. Each mode has its own requirements and features, and is adapted to serve the specific demands of freight and passenger traffic. This gives rise to marked differences in the ways the modes are deployed and utilized in different parts of the world. Recently, there is a trend towards integrating the modes through intermodality and linking the modes ever more closely into production and distribution activities. At the same time, however, passenger and freight activity is becoming increasingly separated across most modes."

Road Transportation

Transportation using road networks are the type of transportation that are connected with movements on constructed roads, carrying people and goods from one place to another on the means of transportation like lorries, cars etc.

Rail Transportation

Transportation by use of rail, restricted to where rails have been built.

Maritime Transportation

Transportation over water, the slowest current form in the movement of goods/people.

Problems with Transportation Geography

Traffic and transportation in existing streets and highways and rail facilities no longer match the new demands created by recent population growth and new location patterns of economic activity. Besides increase in population, another problem is private automobiles overloading the network of highways and arterial streets.

Traffic Congestion

Traffic congestion is a condition on road networks that occurs as use increases, and is characterized by slower speeds, longer trip times, and increased vehicular queueing. The most common example is the physical use of roads by vehicles. When traffic demand is great enough that the interaction between vehicles slows the speed of the traffic stream, congestion is incurred. As demand approaches the capacity of a road (or of the intersections along the road), extreme traffic congestion sets in. When vehicles are fully stopped for periods of time, this is colloquially known as a traffic jam.

Causes

Traffic congestion occurs when a volume of traffic or modal split generates demand for space greater than the available road capacity, this is point is commonly termed saturation. There are a number of specific circumstances which cause or aggravate congestion; most of them reduce the capacity of a road at a given point or over a certain length, or increase the number of vehicles required for a given volume of people or goods. About half of U.S. traffic congestion is recurring, and is attributed to sheer weight of traffic; most of the rest is attributed to traffic incidents, road works and weather events. Speed and flow can also affect network capacity though the relationship is complex.

Traffic research still cannot fully predict under which conditions a "traffic jam" (as opposed to heavy, but smoothly flowing traffic) may suddenly occur. It has been found that individual incidents (such as accidents or even a single car braking heavily in a previously smooth flow) may cause ripple effects (a cascading failure) which then spread out and create a sustained traffic jam when, otherwise, normal flow might have continued for some time longer.

Mathematical Theories

Some traffic engineers have attempted to apply the rules of fluid dynamics to traffic flow, likening it to the flow of a fluid in a pipe. Congestion simulations and real-time observations have shown that in heavy but free flowing traffic, jams can arise spontaneously, triggered by minor events ("butterfly effects"), such as an abrupt steering manoeuvre by a single motorist. Traffic scientists liken such a situation to the sudden freezing of supercooled fluid. However, unlike a fluid, traffic flow is often affected by signals or other events at junctions that periodically affect the smooth flow of traffic. Alternative mathematical theories exist, such as Boris Kerner's three phase traffic theory.

Because of the poor correlation of theoretical models to actual observed traffic flows, transportation planners and highway engineers attempt to forecast traffic flow using empirical models. Their working traffic models typically use a combination of macro-, micro-and mesoscopic features, and may add matrix entropy effects, by "platooning" groups of vehicles and by randomising the flow patterns within individual segments of the network. These models are then typically calibrated by measuring actual traffic flows on the links in the network, and the baseline flows are adjusted accordingly.

It is now claimed that equations can predict these in detail:

> *Phantom jams can form when there is a heavy volume of cars on the road. In that high density of traffic, small disturbances (a driver hitting the brake too hard, or getting too close to another car) can quickly become amplified into a full-blown, self-sustaining traffic jam...*

A team of MIT mathematicians has developed a model that describes how and under what conditions such jams form, which could help road designers minimize the odds of their formation. The researchers reported their findings May 26 in the online edition of Physical Review E.

Key to the new study is the realization that the mathematics of such jams, which the researchers call 'jamitons,' are strikingly similar to the equations that describe detonation waves produced by explosions, says Aslan Kasimov, lecturer in MIT's Department of Mathematics. That discovery enabled the team to solve traffic jam equations that were first theorized in the 1950s.

Economic Theories

Congested roads can be seen as an example of the tragedy of the commons. Because roads in most places are free at the point of usage, there is little financial incentive for drivers not to over-use them, up to the point where traffic collapses into a jam, when demand becomes limited by opportunity cost. Privatization of highways and road pricing have both been proposed as measures that may reduce congestion through economic incentives and disincentives. Congestion can also happen due to non-recurring highway incidents, such as a crash or roadworks, which may reduce the road's capacity below normal levels.

Economist Anthony Downs, in his books *Stuck in Traffic* (1992) and *Still Stuck in Traffic* (2004), argues that rush hour traffic congestion is inevitable because of the benefits of having a relatively standard work day. In a capitalist economy, goods can be allocated either by pricing (ability to pay) or by queueing (first-come first-serve); congestion is an example of the latter.

Instead of the traditional solution of making the "pipe" large enough to accommodate the total demand for peak-hour vehicle travel (a supply-side solution), either by widening roadways or increasing "flow pressure" via automated highway systems, Downs advocates greater use of road pricing to reduce congestion (a demand-side solution, effectively rationing demand), in turn plowing the revenues generated there from into public transportation projects. Road pricing itself is controversial, more information is available in the dedicated article.

Classification

Qualitative classification of traffic is often done in the form of a six letter A-F level of service (LOS) scale defined in the Highway Capacity Manual, a US document used (or used as a basis for national guidelines) worldwide. These levels are used by transportation engineers as a shorthand and to describe traffic levels to the lay public. While this system generally uses delay as the basis for its measurements, the particular measurements and statistical methods vary depending on the facility being described. For instance, while the percent time spent following a slower-moving vehicle figures into the LOS for a rural two-lane road, the LOS at an urban intersection incorporates such measurements as the number of drivers forced to wait through more than one signal cycle.

Negative Impacts

Traffic congestion has a number of negative effects:

- Wasting time of motorists and passengers ("opportunity cost"). As a non-productive activity for most people, congestion reduces regional economic health.
- Delays, which may result in late arrival for employment, meetings, and education, resulting in lost business, disciplinary action or other personal losses.
- Inability to forecast travel time accurately, leading to drivers allocating more time to travel "just in case", and less time on productive activities.
- Wasted fuel increasing air pollution and carbon dioxide emissions owing to increased idling, acceleration and braking. Increased fuel use may also in theory cause a rise in fuel costs.
- Wear and tear on vehicles as a result of idling in traffic and frequent acceleration and braking, leading to more frequent repairs and replacements.
- Stressed and frustrated motorists, encouraging road rage and reduced health of motorists
- Emergencies: blocked traffic may interfere with the passage of emergency vehicles travelling to their destinations where they are urgently needed.
- Spillover effect from congested main arteries to secondary roads and side streets as alternative routes are attempted ('rat running'), which may affect neighbourhood amenity and real estate prices.

Countermeasures

It has been suggested by some commentators that the level of congestion that society tolerates is a rational (though not necessarily conscious) choice between the costs of improving the transportation system (in infrastructure or management) and the benefits of quicker travel. Others link it largely to subjective lifestyle choices, differentiating between car-owning and car-free households.

Road infrastructure:

- Junction improvements

- o Grade separation, using bridges (or, less often, tunnels) freeing movements from having to stop for other crossing movements
- o Ramp signalling, 'drip-feeding' merging traffic via traffic signals onto a congested motorway-type roadway
- o Reducing junctions
  - * Local-express lanes, providing through lanes that bypass junction on-ramp and off-ramp zones
  - * Limited-access road, roads that limit the type and amounts of driveways along their lengths

- Reversible lanes, where certain sections of highway operate in the opposite direction on different times of the day/days of the week, to match asymmetric demand. This may be controlled by Variable-message signs or by movable physical separation
- Separate lanes for specific user groups (usually with the goal of higher people throughput with fewer vehicles)
  - o Bus lanes as part of a busway system
  - o HOV lanes, for vehicles with at least three (sometimes at least two) riders, intended to encourage carpooling
    - * Slugging, impromptu carpooling at HOV access points, on a hitchhiking or payment basis
    - * Market-based carpooling with pre-negotiated financial incentives for the driver

Urban Planning and Design

City planning and urban design practices can have a huge impact on levels of future traffic congestion, though they are of limited relevance for short-term change.

- Grid plans including Fused Grid road network geometry, rather than tree-like network topology which branches into cul-de-sacs (which reduce local traffic, but increase total distances driven and discourage walking by reducing connectivity). This avoids concentration of traffic on a small number of arterial roads and allows more trips to be made without a car.
- Zoning laws that encourage mixed-use development, which reduces distances between residential, commercial, retail, and recreational destinations (and encourage cycling and walking)
- Carfree cities, car-light cities, and eco-cities designed to eliminate the need to travel by car for most inhabitants.

- Transit-oriented development are residential and commercial areas designed to maximize access to public transport.

Supply and Demand

Congestion can be reduced by either increasing road capacity (supply), or by reducing traffic (demand). Capacity can be increased in a number of ways, but needs to take account of latent demand otherwise it may be used more strongly than anticipated. Critics of the approach of adding capacity have compared it to "fighting obesity by letting out your belt" (inducing demand that did not exist before).

Reducing road capacity has in turn been attacked as removing free choice as well as increasing travel costs and times.

Increased supply can include:

- Adding more capacity at bottlenecks (such as by adding more lanes at the expense of hard shoulders or safety zones, or by removing local obstacles like bridge supports and widening tunnels)
- Adding more capacity over the whole of a route (generally by adding more lanes)
- Creating new routes
- Traffic management improvements.

Reduction of demand can include:

- Parking restrictions, making motor vehicle use less attractive by increasing the monetary and non-monetary costs of parking, introducing greater competition for limited city or road space. Most transport planning experts agree that free parking distorts the market in favour of car travel, exacerbating congestion.
- Park and ride facilities allowing parking at a distance and allowing continuation by public transport or ride sharing. Park-and-ride car parks are commonly found at metro stations, freeway entrances in suburban areas, and at the edge of smaller cities.
- Reduction of road capacity to force traffic onto other travel modes. Methods include traffic calming and the shared space concept.
- Road pricing, charging money for access onto a road/specific area at certain times, congestion levels or for certain road users.
    - o "Cap and trade", in which only licensed cars are allowed on the roads. A limited quota of car licences are issued each year and traded in a free market fashion. This guarantees

that the number of cars does not exceed road capacity while avoiding the negative effects of shortages normally associated with quotas. However since demand for cars tends to be inelastic, the result are exorbitant purchase prices for the licenses, pricing out the lower levels of society, as seen Singapore's Certificate of Entitlement scheme.

- o Congestion pricing, where a certain area, such as the inner part of a congested city, is surrounded with a cordon into which entry with a car requires payment. The cordon may be a physical boundary (i.e., surrounded by toll stations) or it may be virtual, with enforcement being via spot checks or cameras on the entry routes. Major examples are Singapore's electronic road pricing, the London congestion charge system, Stockholm congestion tax and the use of HOT lanes predominately in North America.

• Road space rationing, where regulatory restrictions prevent certain types of vehicles from driving under certain circumstances or in certain areas.

- o Number plate restrictions based on days of the week, as practiced in several large cities in the world, such as Athens, Mexico City and Sao Paulo. In effect, such cities are banning a different part of the automobile fleet from roads each day of the week. Mainly introduced to combat smog, these measures also reduce congestion. A weakness of this method is that richer drivers can purchase a second or third car to circumvent the ban.
- o Permits, where only certain types of vehicles (such as residents) are permitted to enter a certain area, and other types (such as through-traffic) are banned. For example, Bertrand Delanoe, the mayor of Paris, has proposed to impose a complete ban on motor vehicles in the city's inner districts, with exemptions only for residents, businesses, and the disabled.

• Policy approaches, which usually attempt to provide either strategic alternatives or which encourage greater usage of existing alternatives through promotion, subsidies or restrictions.

- o Incentives to use public transport, increasing modal shares. This can be achieved through infrastructure investment, subsidies, transport integration, pricing strategies that decrease the marginal cost/fixed cost ratios, improved timetabling and greater priority for buses to reduce journey time e.g. [Bus Lanes], [BTR].

- o Cycling promotion through legislation, cycle facilities, subsidies, and awareness campaigns. The Netherlands has been pursuing cycle friendly policies for decades, and around a quarter of their commuting is done by bicycle.
- o Telecommuting encouraged through legislation and subsidies.
- o Online shopping promotion, potentially with automated delivery booths helping to solve the last mile problem and reduce shopping trips made by car.

Traffic Management

Use of so-called Intelligent transportation system, which guide traffic:

- Traffic reporting, via radio, GPS or possibly mobile phones, to advise road users
- Variable message signs installed along the roadway, to advise road users
- Navigation systems, possibly linked up to automatic traffic reporting
- Traffic counters permanently installed, to provide real-time traffic counts
- Convergence indexing road traffic monitoring, to provide information on the use of highway on-ramps
- Automated highway systems, a future idea which could reduce the safe interval between cars (required for braking in emergencies) and increase highway capacity by as much as 100% while increasing travel speeds
- Parking guidance and information systems providing dynamic advice to motorists about free parking
- Active Traffic Management system opens up UK motorway hard shoulder as an extra traffic lane, it uses CCTV and VMS to control and monitor the traffic's use of the extra lane

Other associated:

- School opening times arranged to avoid rush hour traffic (in some countries, private car school pickup and drop-off traffic are substantial percentages of peak hour traffic).

- Considerate driving behaviour promotion and enforcement. Driving practices such as tailgating and frequent lane changes can reduce a road's capacity and exacerbate jams. In some countries signs are placed on highways to raise awareness, while others have introduced legislation against inconsiderate driving.
- Visual barriers to prevent drivers from slowing down out of curiosity (often called "rubbernecking" in the United States). This often includes accidents, with traffic slowing down even on roadsides physically separated from the crash location. This also tends to occur at construction sites, which is why some countries have introduced rules that motorway construction has to occur behind visual barrier
- Speed limit reductions, as practiced on the M25 motorway in London. With lower speeds allowing cars to drive closer together, this increases the capacity of a road. Note that this measure is only effective if the interval between cars is reduced, not the distance itself. Low intervals are generally only safe at low speeds.
- Lane splitting/filtering, where space-efficient vehicles, usually motorcycles, scooters, and ultra-narrow cars ride or drive in the space between cars, buses, and trucks. This is however illegal in many countries as it is perceived as a safety risk.

By Country

Iran

Because of low price of gas and gasoil in Iran and inadequate public transportation traffic congestion is a conmen problem in different cities like Mash'had, Isfahan, Shiraz and especially Tehran (capital city of Iran). Recently developing Metro and BRT systems in Tehran and strategies for limiting gas uses has been applied to reduce car using, but unfortunately the problem is still crucial.

Australia

Traffic during peak hours in major Australian cities, such as Melbourne, Sydney, Brisbane and Perth, is usually very congested and can cause considerable delay for motorists. Australians rely mainly on radio and television to obtain current traffic information. GPS, webcams, and online resources are increasingly being used to monitor and relay traffic conditions to motorists. Measures put in place by the federal and state government to combat traffic congestion include construction of new road infrastructure and increased investment in public transport. In Brisbane, ongoing road works projects on many major roads have caused ongoing congestion throughout the city and increased commutes considerably.

Brazil

In Brazil the recent records of traffic jams over the major big cities are recognized by public authorities as one of the main challenges for Sao Paulo, Rio de Janeiro, Belo Horizonte, Brasilia, Curitiba and Porto Alegre, where due to the country's economic bonanza, the automobile fleets have almost doubled in several of these cities from 2000 to 2008. According to Time Magazine, Sao Paulo has the world's worst traffic jams. On June 10, 2009, the historical record was set with more than 182 miles (293 km) of accumulated queues out of 522 mi (835 km) being monitored. Despite implementation since 1997 of road space rationing by the last digit of the plate number during rush hours every weekday, traffic in this 20 million city still experiences severe congestion. According to experts, this is due to the accelerated rate of motorization occurring since 2003, in Sao Paulo the fleet is growing at a rate of 7.5% per year, with almost 1,000 new cars bought in the city every day, and the limited capacity of public transport. The subway has only 38 miles (61 km) of lines, though 22 further miles are under construction or planned by 2010. Every day, many citizens spend between three up to four hours behind the wheel. In order to mitigate the aggravating congestion problem, since June 30, 2008 the road space rationing program was expanded to include and restrict trucks and light commercial vehicles.

Colombia

In Bogota the excessive traffic jams cause high levels of stress in people, and are the main cause of air pollution. The problem has been mitigated partially since 2000 through the implementation of the Trans Milenio, a bus rapid transit system that has been improving mobility throughout the city. The city also restricts use of vehicles several days each week depending on the last digits of license plates. However, this system, called 'Pico y placa' tends to promote the purchase of second cars by the wealthy.

Hong Kong

Hong Kong aborted a congestion pricing system in the 1980s due to public pressure and has since relied on a vehicle high purchase tax to discourage overall car purchasing but has developed no localised congestion management techniques.

Netherlands

The road network in the Netherlands is usually congested in the morning and afternoon rush hour on working days. However, the rush hour periods seem to have become longer and longer and one may occasionally run into congestion any time of the day or night. Commuter traffic to and from major cities such as Amsterdam, Rotterdam, The Hague, Utrecht, Eindhoven, Zwolle, Enschede and Groningen may cause congestion. Congestion is difficult to resolve because the Netherlands is a relatively densely populated country where there is little room for expansion. Proposals for a "pay per distance travelled" is thought to discourage car driving but has not been implemented yet due to car owner resistance.

New Zealand

New Zealand has followed strongly car-oriented transport policies since after World War II (especially in the Auckland area, where about one third of the country's population lives), and currently has one of the highest car-ownership rates per capita in the world, after the United States. Because of the negative results, congestion in the big centres is a major problem. Current measures include both the construction of new road infrastructure as well as increased investment in public transport, which had strongly declined in all cities of the country except Wellington.

United Kingdom

In the United Kingdom the inevitability of congestion in some urban road networks has been officially recognised since the Department for Transport set down policies based on the report *Traffic in Towns* in 1963:

> *Even when everything that it is possibly to do by way of building new roads and expanding public transport has been done, there would still be, in the absence of deliberate limitation, more cars trying to move into, or within our cities than could possibly be accommodated..*

The Department for Transport sees growing congestion as one of the most serious transport problems facing the UK. On 1 December 2006, Rod Eddington published a UK government-sponsored report into the future of Britain's transport infrastructure. The Eddington

Transport Study set out the case for action to improve road and rail networks, as a "crucial enabler of sustained productivity and competitiveness". Eddington has estimated that congestion may cost the economy of England £22 bn a year in lost time by 2025.

He warned that roads were in serious danger of becoming so congested that the economy would suffer. At the launch of the report Eddington told journalists and transport industry representatives introducing road pricing to encourage drivers to drive less was an "economic no-brainer". There was, he said "no attractive alternative". It would allegedly cut congestion by half by 2025, and bring benefits to the British economy totalling £28 bn a year.

United States

The Texas Transportation Institute estimated that, in 2000, the 75 largest metropolitan areas experienced 3.6 billion vehicle-hours of delay, resulting in 5.7 billion U.S. gallons (21.6 billion litres) in wasted fuel and $67.5 billion in lost productivity, or about 0.7% of the nation's GDP.

It also estimated that the annual cost of congestion for each driver was approximately $1,000 in very large cities and $200 in small cities. Traffic congestion is increasing in major cities and delays are becoming more frequent in smaller cities and rural areas.

In 2005, the three areas in the United States with the highest levels of traffic congestion were Los Angeles, New York City, and Chicago. The congestion cost for the Los Angeles area alone was estimated at US$9.325 billion.

Between 1980 and 1999 the total number of miles of vehicle travel increased by 76 percent. National and local highway construction programs have accommodated some, but not all, of this traffic growth.

Venezuela

While most of the world is troubled with high gas prices, Venezuela has the lowest gas price in the world.

They pay 0.097 bolivars, an equivalent of $0.03 per liter or $0.12 per gallon.

Venezuela has fixed their price of gasoline at this rate since 1998, even though it is estimated that the government could save $3 billion dollars a year by cutting 30 minutes from the average drive time.

> *Zarhay Infante leaves home shortly after 5am on a 30km (19 miles) drive to her job in the capital. If her*

> *journey goes well, she gets there three-and-a-half hours later. Three years ago she could go to Caracas in 45 minutes on the motorway. According to Zarhay, "It gets worse every day."*

The government which preceded leftist Hugo Chavez began raising gasoline prices. However, upon taking office Chavez froze prices in bolivar terms.

Nature, Scope and Significance of Transport Geography

"The ideal transport mode would be instantaneous, free, have an unlimited capacity and always be available. It would render space obsolete. This is obviously not the case. Space is a constraint for the construction of transport networks. Transportation appears to be an economic activity different from the others. It trades space with time and thus money" (translated from Merlin, 1992).

As the above quotation underlines, the unique purpose of transportation is to overcome space, which is shaped by a variety of human and physical constraints such as distance, time, administrative divisions and topography. Jointly, they confer a friction to any movement, commonly known as the friction of distance. However, these constraints and the friction they create can only be partially circumscribed.

The extent to which this is done has a cost that varies greatly according to factors such as the distance involved and the nature of what is being transported. There would be no transportation without geography and there would be no geography without transportation. The goal of transportation is thus to transform the geographical attributes of freight, people or information, from an origin to a destination, conferring them an added value in the process. The convenience at which this can be done-transportability-varies considerably.

Transportability refers to the ease of movement of passengers, freight or information. It is related to transport costs as well as to the attributes of what is being transported (fragility, perishable, price). Political factors can also influence transportability such as laws, regulations, borders and tariffs. When transportability is high, activities are less constrained by distance.

The specific purpose of transportation is to fulfil a demand for mobility, since transportation can only exists if it moves people, freight and information around. Otherwise it has no purpose. This is because transportation is dominantly the outcome of a derived demand. Distance, a core attribute of transportation can be represented in a variety of ways, ranging from a simple Euclidean distance-a straight line between two locations-to what can be called logistical distance; a complete set of tasks required to be done so that distance can be overcome. Any movement must thus consider its geographical setting which in turn is linked to spatial flows and their patterns. Urbanization, multinational corporations, the globalization of trade and the international division of labour are all forces shaping and taking advantage of transportation at different, but often related, scales. Globalization involves its own space of flows.

Consequently, the fundamental purpose of transport is geographic in nature, because it facilitates movements between different locations. Transport thus plays a role in the structure and organization of space and territories, which may vary according to the level of development. In the 19th century, the purpose of the emerging modern forms of transportation, mainly railways and maritime shipping, was to expand coverage, and create and consolidate national markets. In the 20th century, the objective shifted to selecting itineraries, prioritizing transport modes, increasing the capacity of existing networks and responding to the mobility needs and this at a scale which was increasingly global. In the 21st century, transportation must cope with a globally oriented economic system in a timely and cost effective way, but also with several local problems such as congestion and capacity constraints.

The Importance of Transportation

Transport represents one of the most important human activities worldwide. It is an indispensable component of the economy and plays a major role in spatial relations between locations. Transport creates valuable links between regions and economic activities, between people and the rest of the world. Transport is a multidimensional activity whose importance is:

- Historical. Transport modes have played several different historical roles in the rise of civilizations (Egypt, Rome and China), in the development of societies (creation of social structures) and also in national defence (Roman Empire, American road network).
- Social. Transport modes facilitate access to healthcare, welfare, and cultural or artistic events, thus performing a social service. They shape social interactions by favouring or inhibiting the

mobility of people. Transportation thus support and may even shape social structures.

- Political. Governments play a critical role in transport as sources of investment and as regulators. The political role of transportation is undeniable as governments often subsidize the mobility of their populations (highways, public transit, etc.). While most transport demand relates to economic imperatives, many communication corridors have been constructed for political reasons such as national accessibility or job creation. Transport thus has an impact on nation building and national unity, but it is also a political tool.
- Economic. The evolution of transport has always been linked to economic development. It is an industry in its own right (car manufacturing, air transport companies, etc.). The transport sector is also an economic factor in the production of goods and services. It contributes to the value-added of economic activities, facilitates economies of scale, influences land (real estate) value and the geographic specialization of regions. Transport is both a factor shaping economic activities, and is also shaped by them.
- Environmental. Despite the manifest advantages of transport, its environmental consequences are also significant. They include air and water quality, noise level and public health. All decisions relating to transport need to be evaluated taking into account the corresponding environmental costs. Transport is a dominant factor in contemporary environmental issues.

Substantial empirical evidence indicates that the importance of transportation is growing. The following contemporary trends can be identified regarding this issue:

- Growth of the demand. The last 50 years have seen a considerable growth of the transport demand related to individual (passengers) as well as freight mobility. This growth is jointly the result of larger quantities of passengers and freight being moved, but also the longer distances over which they are carried. Recent trends underline an ongoing process of mobility growth, which has led to the multiplication of the number of journeys involving a wide variety of modes that service transport demands.
- Reduction of costs. Even if several transportation modes are very expensive to own and operate (ships and planes for

instance), costs per unit transported have dropped significantly over the last decades. This has made it possible to overcome larger distances and further exploit the comparative advantages of space. As a result, despite the lower costs, the share of transport activities in the economy has remained relatively constant in time.

- Expansion of infrastructures. The above two trends have obviously extended the requirements for transport infrastructures both quantitatively and qualitatively. Roads, harbours, airports, telecommunication facilities and pipelines have expanded considerably to service new areas and adding capacity to existing networks. Transportation infrastructures are thus a major component of the land use, notably in developed countries.

Facing these contemporary trends, an important part of the spatial differentiation of the economy is related to where resources (raw materials, capital, people, information etc.) are located and how well they can be distributed. Transport routes are established to distribute resources between places where they are abundant and places where they are scarce, but only if the costs are lower than the benefits.

Consequently, transportation has an important role to play in the conditions that affect global, national and regional economic entities. It is a strategic infrastructure that is so embedded in the socio-economic life of individuals, institutions and corporations that it is often invisible to the consumer, but always part of all economic and social functions. This is paradoxical, since the perceived invisibility of transportation is derived from its efficiency. If transport is disrupted or ceases to operate, the consequences can be dramatic.

Transportation in Geography

Transportation interests geographers for two main reasons. First, transport infrastructures, terminals, equipment and networks occupy an important place in space and constitute the basis of a complex spatial system. Second, since geography seeks to explain spatial relationships, transport networks are of specific interest because they are the main support of these interactions.

Transport geography is a sub-discipline of geography concerned about movements of freight, people and information. It seeks to link spatial constraints and attributes with the origin, the destination, the extent, the nature and the purpose of movements.

Transport geography, as a discipline, emerged from economic geography in the second half of the twentieth century. Traditionally, transportation has been an important factor behind the economic representations of the geographic space, namely in terms of the location of economic activities and the monetary costs of distance. The growing mobility of passengers and freight justified the emergence of transport geography as a specialized field of investigation. In the 1960s, transport costs were recognized as key factors in location theories and transport geography began to rely increasingly on quantitative methods, particularly over network and spatial interactions analysis. However, from the 1970s globalization challenged the centrality of transportation in many geographical and regional development investigations. As a result, transportation became under represented in economic geography in the 1970s and 1980s, even if mobility of people and freight and low transport costs were considered as important factors behind the globalization of trade and production.

Since the 1990s, transport geography has received renewed attention, especially because the issues of mobility, production and distribution are interrelated in a complex geographical setting. It is now recognized that transportation is a system that considers the complex relationships between its core elements. These core elements are networks, nodes and demand. Transport geography must be systematic as one element of the transport system is linked with numerous others. An approach to transportation thus involves several fields where some are at the core of transport geography while others are more peripheral. However, three central concepts to transport systems can be identified:

- Transportation nodes. Transportation primarily links locations, often characterized as nodes. They serve as access points to a distribution system or as transshipment/intermediary locations within a transport network. This function is mainly serviced by transport terminals where flows originate, end or are being transshipped from one mode to the other. Transport geography must consider its places of convergence and transshipment.
- Transportation networks. Considers the spatial structure and organization of transport infrastructures and terminals. Transport geography must include in its investigation the infrastructures supporting and shaping movements.
- Transportation demand. Considers the demand for transport services as well as the modes used to support movements.

Once this demand is realized, it becomes an interaction which flows through a transport network. Transport geography must evaluate the factors affecting its derived demand function.

The analysis of these concepts relies on methodologies often developed by other disciplines such as economics, mathematics, planning and demography. Each provides a different dimension to transport geography. For instance, the spatial structure of transportation networks can be analyzed with graph theory, which was initially developed for mathematics. Further, many models developed for the analysis of movements, such as the gravity model, were borrowed from physical sciences. Multi disciplinarity is consequently an important attribute of transport geography, as in geography in general.

The role of transport geography is to understand the spatial relations that are produced by transport systems. This gives rise to several fallacies about transportation. A better understanding of spatial relations is essential to assist private and public actors involved in transportation mitigate transport problems, such as capacity, transfer, reliability and integration of transport systems. There are three basic geographical considerations relevant to transport geography:

- Location. As all activities are located somewhere, each location has its own characteristics conferring a potential supply and/ or a demand for resources, products, services or labour. A location will determine the nature, the origin, the destination, the distance and even the possibility of a movement to be realized. For instance, a city provides employment in various sectors of activity in addition to consume resources.
- Complementarity. Locations must require exchanging goods, people or information. This implies that some locations have a surplus while others have a deficit. The only way an equilibrium can reached is by movements between locations having surpluses and locations having demands. For instance, a complementarity is created between a store (surplus of goods) and its customers (demand of goods).
- Scale. Movements generated by complementarity are occurring at different scales, pending the nature of the activity. Scale illustrates how transportation systems are established over local, regional and global geographies. For instance, home-to-work journeys generally have a local or regional scale, while the distribution network of a multinational corporation is most likely to cover several regions of the world.

Consequently, transport systems, by their nature, consume land and support the relationships between locations.

Sociopolitical Implications of Transport Geography

It may seem axiomatic to argue that transportation is a necessary, but not sufficient, component of growth and development.

Decades of research have exposed the critical relationships between infrastructure, accessibility, mobility, policy, and social change.

Simply put, those regions and places that are better endowed with transportation have fared better overall, as measured by macroeconomic statistics of development, than those that are poorly equipped. Simple comparisons of countries like Haiti, Afghanistan, or Chad with Japan, Germany, or the United States suggest that the role of transportation in driving socio-economic change is critical.

Yet transportation is more than just the provision of infrastructure, facilities, networks, or investment; it is inextricably intertwined with how humans interact through policies, ideologies, and societies across time and space.

Transportation provides a fundamental foundation for the building blocks of societies – labour, capital, territory – and intersects with the human and physical environment in ways that have profound geographical consequences.

Progress in transportation geography research has been impressive over the nearly two decades since Rimmer (1988) completed a series of reports on the state of the subdiscipline. New research theories and methodologies have been stimulated, in part, by the growing importance of globalization as both ideology and process, and the evolution of spatial analytical technologies. The notion that one can now make anything anywhere on the planet and sell anything anywhere on the planet (political and economic barriers notwithstanding) argues for an intensity of interaction between people, goods, and information that has motivated significant shifts in the way that accessibility is analyzed at multiple scales. More sophisticated analytical and computing capabilities (GIS, for example) have facilitated broader, deeper, and more interrelated approaches to transportation research.

One-dimensional, structural approaches to transportation have been superseded by research agendas that embrace myriad perspectives on the relationship between transport and society. Yet significant research challenges remain, not least of which is the need to provide a more scaled and integrative approach to transportation's relationship with people and places. In this first of three progress reports, my focus is on transportation research at the global scale. In future reports, transportation research at both the regional and local contexts will be examined. Before commencing this journey, however, a brief discussion of the broad framework of transportation research provides some context to how place and space are currently engaged with by transport geographers. Over the course of the 20th century, the automobile rapidly developed from an expensive toy for the rich into the *de facto* standard for passenger transport in most developed countries. In developing countries, the effects of the automobile have lagged, but are emulating the impacts of developed nations. The development of the automobile built upon the transport revolution started by railways, and like the railways, introduced sweeping changes in employment patterns, social interactions, infrastructure and goods distribution.

The effects of the automobile on everyday life have been a subject of controversy. While the introduction of the mass-produced automobile represented a revolution in mobility and convenience, the modern consequences of heavy automotive use contribute to the use of non-renewable fuels, a dramatic increase in the rate of accidental death, social isolation and the disconnection of community, rise in obesity, the generation of air and noise pollution, and the facilitation of urban sprawl and urban decay.

Access and Convenience

Worldwide the automobile has allowed easier access to remote places. However, average journey times to regularly visited places have increased in large cities, especially in Latin America, as a result of widespread automobile adoption. This is due to traffic congestion and the increased distances between home and work brought about by urban sprawl.

Examples of automobile access issues in underdeveloped countries are:

- Paving of the Mexico Pacific Coast highway through Baja California, completing the connection of Cabo San Lucas to California, allowing the first routine travel along that route and the first convenient access to the outside world for villagers along the route.
- In Madagascar, approximately 30 percent of the population does not have access to reliable all weather roads.

- In China, there are currently 184 towns and 54,000 villages that have no access to automobile use (or roads at all)
- The origin of HIV explosion in the human population has been hypothesized by CDC researchers to derive in part from more intensive social interactions afforded by new road networks in Central Africa allowing more frequent travel from villagers to cities and higher density development of many African cities in the period 1950 to 1980.

The following developments in retail are partially due to automobile use:

- Drive-thru fast food purchasing
- Gasoline station grocery shopping.

Economic Changes

The development of the automobile has contributed to changes in employment distribution, shopping patterns, social interactions, manufacturing priorities and city planning; increasing use of automobiles has reduced the roles of walking, horses and railroads.

Infrastructure

Aside from industries, one of the most visible effects the automobile has had on the world is the huge increase in the amount of surfaced roads. For example, between 1921 and 1941, the United States spent US$40 billion on roads, increasing the amount of surfaced road from 387,000 miles (619,000 kilometres) to over 1,000,000 miles (1.6 million kilometres) which does not even take into account road widening.

United States

In addition to federal, state, and local dollars for roadway construction, car use was also encouraged through new zoning laws that required that any new business construct a certain amount of parking based on the size and type of facility. The effect of this was to create a massive quantity of free parking spaces and to push businesses further back from the road. Many shopping centres and suburbs abandoned sidewalks altogether, making pedestrian access dangerous. This had the effect of encouraging people to drive, even for short trips that might have been walkable, thus increasing and solidifying American auto-dependency. As a result of this change, employment opportunities for people who were not wealthy enough to own a car and for people who could not drive, due to age or physical disabilities, became severely limited.

Environmental Impact

For much of the early history of the car, no consideration was given to various environmental effects caused by the automobile. Automobiles are a major source of air pollution and noise pollution. The manufacture and use of automobiles makes up 20 to 25 percent of the carbon dioxide emissions that are believed to be causing global climate change. There are over 600 million cars and light vehicles (excluding heavy trucks and buses) worldwide, The automobile contributes significantly to noise pollution worldwide; in response to these impacts, an entire technology of noise barrier design and other noise mitigation has emerged. In the United States the typical car emits approximately 3.4 grams per mile of carbon monoxide.

With increased road-building came negative effects on habitat for wildlife, primarily through habitat fragmentation and surface runoff alteration. New roads built through sensitive habitat can cause the loss or degradation of ecosystems, and the materials required for roads come from large-scale rock quarrying and gravel extraction, which sometimes occurs in sensitive ecological areas. Road construction also alters the water table, increases surface runoff, and increases the risk of flooding.

Cultural Changes

Prior to the appearance of the automobile, horses, walking and streetcars were the major modes of transportation within cities. Horses require a large amount of care, and were therefore kept in public facilities that were usually far from residences. The manure they left on the streets also created a sanitation problem.

The automobile made regular medium-distance travel more convenient and affordable, especially in areas without railways. Because automobiles did not require rest, and were faster than horse-drawn conveyances, people were routinely able to travel farther than in earlier times. The construction of highways half a century later continued this revolution in mobility. Some experts suggest that many of these changes began during the *Golden age of the bicycle*, the preceding era from 1880—1915.

Changes to Urban Society

Beginning in the 1940s, most urban environments in the United States lost their streetcars, cable cars, and other forms of light rail, to be replaced by diesel-burning motor coaches or buses. Many of these have never returned, though some urban communities eventually installed subways. Another change brought about by the automobile is that modern urban pedestrians must be more alert than their ancestors. In the past, a pedestrian had to worry about relatively slow-moving streetcars or other obstacles of travel. With the proliferation of the automobile, a pedestrian has to anticipate safety risks of automobiles at high speeds because cars may cause serious damage to a human.

According to many social scientists, the loss of pedestrian-scale villages has also disconnected communities. Many people in developed countries have less contact with their neighbours and rarely walk unless they place a high value on exercise.

Advent of Suburban Society

Because of the automobile, the outward growth of cities accelerated, and the development of suburbs in automobile intensive cultures was intensified. Until the advent of the automobile, factory workers lived either close to the factory or in high density communities farther away, connected to the factory by streetcar or rail. The automobile and the federal subsidies for roads and suburban development that supported car culture allowed people to live in low density communities far from the city centre and integrated city neighbourhoods. The developing suburbs created few local jobs, due to single use zoning. Hence, residents commuted longer distances to work each day as the suburbs expanded.

Car Culture

The car had a significant effect on the culture of the middle class. Automobiles were incorporated into all parts of life from music to books to movies. Between 1905 and 1908, more than 120 songs were written in which the automobile was the subject. The automotive themes of these songs reflected the general culture of the automotive industry: sexual adventure, liberation from social control, and masculine power. Books centered on motor boys who liberated themselves from the average, normal, middle class life, to travel and seek adventure in the exotic. Car ownership came to be associated with independence, freedom, and increased status.

George Monbiot writes that widespread car culture has shifted voter's preference to the right of the political spectrum. He thinks that car culture has contributed to an increase in individualism and fewer social interactions between members of different socioeconomic classes. Since the early days of the automobile, car manufacturers and petroleum fuel suppliers successfully lobbied governments to build public roads. Road building was sometimes also influenced by Keynesian-style political ideologies. In Europe, massive freeway building programs were initiated by a number of social democratic governments after World War II, in an attempt to create jobs and make the automobile available to the working classes. From the 1970s, promotion of the automobile increasingly became a trait of some conservatives. Margaret Thatcher talked of a "great car economy", and increased government spending on roads.

Safety

Motor vehicle accidents are attributed to 37.5% of accidental deaths in the United States, making them the country's leading cause of accidental death.

Costs

In countries such as the United States the infrastructure that makes car use possible, such as highways, roads and parking lots is funded by the government and supported through government zoning and construction requirements. The gas tax covers about 60% of highway construction and repair costs. Payments by motor-vehicle users fall short of government expenditures tied to motor-vehicle use by 20–70 cents per gallon of gas. Zoning laws in many areas require that large, free parking lots accompany any new buildings.

Municipal parking lots are often free or do not charge a market-rate. Hence, the cost of driving a car in the US is subsidized, supported by businesses and the government who cover the cost of roads and parking.

This government support of the automobile through subsidies for infrastructure, the cost of highway police enforcement, recovering stolen cars, and many other factors makes public transport a less economically competitive choice for commuters when considering direct out-of-pocket costs. Consumers often make choices based on the direct out-of-pocket costs and underestimate the indirect costs of car ownership, auto insurance and car maintenance.

However, globally and in some US cities, tolls and parking fees partially offset these heavy subsidies for driving. Transportation planning policy advocates often support tolls, increased gas taxes, congestion pricing and market-rate pricing for municipal parking as a means of balancing car use in urban centres with more efficient, less environmentally and socially destructive modes of transportation such as buses and trains.

When cities charge market rates for on-street parking and municipal parking garages, and when bridges and tunnels are tolled, driving becomes less competitive in terms of out-of-pocket costs than other modes of transportation. When municipal parking is underpriced and roads are not tolled, most of the cost of vehicle usage is paid for by general government revenue, a subsidy for motor vehicle use. The size of this subsidy dwarfs the federal, state, and local subsidies for the maintenance of infrastructure and discounted fares for public transportation.

By contrast, although there are environmental and social costs for rail, there is a very small impact.

4

Transport Economics

Transport economics is a branch of economics that deals with the allocation of resources within the transport sector and has strong linkages with civil engineering. Transport economics differs from some other branches of economics in that the assumption of a spaceless, instantaneous economy does not hold. People and goods flow over networks at certain speeds. Demands peak. Advanced ticket purchase is often induced by lower fares. The networks themselves may or may not be competitive. A single trip (the final good from the point-of-view of the consumer) may require bundling the services provided by several firms, agencies and modes.

Although transport systems follow the same supply and demand theory as other industries, the complications of network effects and choices between non-similar goods (e.g. car and bus travel) make estimating the demand for transportation facilities difficult. The development of models to estimate the likely choices between the non-similar goods involved in transport decisions (discrete choice models) led to the development of an important branch of econometrics, and a Nobel Prize for Daniel McFadden.

In transport, demand can be measured in numbers of journeys made or in total distance travelled across all journeys (e.g. passenger-kilometres for public transport or vehicle-kilometres of travel (VKT) for private transport). Supply is considered to be a measure of capacity. The price of the good (travel) is measured using the generalised cost of travel, which includes both money and time expenditure. The effect of increases in supply (capacity) are of particular interest in transport economics, as the potential environmental consequences are significant.

Externalities

In addition to providing benefits to their users, transport networks impose both positive and negative externalities on non-users. The consideration of these externalities-particularly the negative ones-is a part of transport economics.

Positive externalities of transport networks may include the ability to provide emergency services, increases in land value and agglomeration benefits. Negative externalities are wide-ranging and may include local air pollution, noise pollution, light pollution, safety hazards, community severance and congestion. The contribution of transport systems to potentially hazardous climate change is a significant negative externality which is difficult to evaluate quantitatively, making it difficult (but not impossible) to include in transport economics-based research and analysis.

Congestion is considered a negative externality by economists. An externality occurs when a transaction causes costs or benefits to third party, often, although not necessarily, from the use of a public good. For example, manufacturing or transportation cause air pollution imposing costs on others when making use of public air.

Traffic Congestion

Traffic congestion is a negative externality caused by various factors. A 2005 American study stated that there are seven root causes of congestion, and gives the following summary of their contributions: bottlenecks 40%, traffic incidents 25%, bad weather 15%, work zones 10%, poor signal timing 5%, and special events/other 5%. Within the transport economics community, congestion pricing is considered to be an appropriate mechanism to deal with this problem (i.e. to internalise the externality) by allocating scarce roadway capacity to users. Capacity expansion is also a potential mechanism to deal with traffic congestion, but is often undesirable (particularly in urban areas) and sometimes has questionable benefits. William Vickrey, winner of the 1996 Nobel Prize for his work on "moral hazard", is considered one of the fathers of congestion pricing, as he first proposed it for the New York City subway system in 1952. In the road transportation arena these theories were extended by Maurice Allais, a fellow Nobel prize winner "for his pioneering contributions to the theory of markets and efficient utilization of resources", Gabriel Roth who was instrumental in the first designs and upon whose World Bank recommendation the first system was put in place in Singapore.

Reuben Smeed, the deputy director of the Transport and Road Research Laboratory was also a pioneer in this field, and his ideas were presented to the British government in what is known as the Smeed Report. Congestion is not limited to road networks; the negative externality imposed by congestion is also important in busy public transport networks as well as crowded pedestrian areas.

Road Pricing

Road pricing is an economic concept regarding the various direct charges applied for the use of roads. The road charges includes fuel taxes, licence fees, parking taxes, tolls, and congestion charges, including those which may vary by time of day, by the specific road, or by the specific vehicle type, being used. Road pricing has two distinct objectives: revenue generation, usually for road infrastructure financing, and congestion pricing for demand management purposes. Toll roads are the typical example of revenue generation. Charges for using high-occupancy toll lanes or urban tolls for entering a restricted area of a city are typical examples of using road pricing for congestion management purposes.

European Application

Facing rising levels of traffic congestion, European governments are giving serious consideration to nationwide road pricing schemes. Some of these could exploit the new Galileo satellite positioning system, although it is possible to arrange road pricing using various different technologies. A satellite based system would entail vehicles containing a satellite tracking device which would determine which roads were being driven along, for how far and at what time of day. This information would then be sent to a central computer system, and the appropriate charges levied against the driver.

Germany

Schemes for charging trucks (lorries) in Germany (by the company Toll Collect) and Austria are already underway. The LKW-MAUT road pricing scheme began on January 1, 2005, trucks pay between €0.09 and €0.14 per kilometre depending on their emission levels and number of axles. The expensive scheme, combining satellite technology with other technologies, suffered numerous delays before implementation, whilst a scheme using much simpler technology in Austria was up and running in 2004.

In the UK, the Labour government announced in July 2005 that the proposed UK truck road user charging scheme would not go ahead.

Italy

A traffic charge program in Milan, called "Ecopass", began on a trial basis on January 2, 2008. It exempts vehicles compliant with the Euro 3 and Euro 4 emission standards or higher, as well as several alternative fuel vehicles. Residents within the restricted zone, called ZTL (Italian: Zone a Traffico Limitato), may purchase a discounted annual pass. Although the program is operationally similar to existing congestion pricing schemes, its main objective is to reduce air pollution from vehicle emissions rather than relieve traffic congestion. The program was extended until December 31, 2009, and a public consultation will be conducted to decide if the charge should become permanent.

Malta

A fully automated system called a Controlled Vehicular Access (CVA) system has been launched in Malta's capital city of Valletta since May 1, 2007. When compared to other countries that make use of congestion charging models, the Maltese system makes use of a wider array of innovations including variable payments according to the duration of stay, flexible exemption rules, including exemptions for residents within the charging zone, and monthly or quarterly billing options for vehicle owners. Pre-payment facilities, including direct debit arrangements and purposely designed vouchers, are also available. The billing system was designed in Malta and has been described as a state of the art 'next generation congestion charge billing solution'. The Valletta Congestion Charge, which is also known as Valletta CVA, was recently nominated for the Best European Transport Strategy Award. Public voting is still underway.

Norway

One of the earliest schemes was introduced in Bergen in Norway in 1986. Only traffic entering the town is charged and only during weekdays from 6:00 a.m. through 10:00 p.m. Public service vehicles pay no charge.

Bergen has now a fully automated toll plaza system that is based on passing without stopping for all traffic. There are no coin slots or manual service. A similar system was introduced for the Oslo Toll Ring from February 2, 2008. To ensure interoperability of electronic fee collection in Norway a system called Auto PASS is used throughout the country for toll roads and congestion charging schemes etc. Most local drivers have purchased a tag which is automatically read on passing the detectors. As of February 2008, there will be six fully automated schemes in operation. Motorists without a tag pay a fee at a manual barrier.

Sweden

Stockholm has a congestion pricing system, Stockholm congestion tax, in use on a permanent basis since August 1, 2007, after having had a seven month trial period from January 3 to July 31, 2006. The City Centre is within the congestion tax zone. All the entrances and exits of this area have unmanned control points operating with automatic number plate recognition. All vehicles entering or exiting the congestion tax affected area, with a few exceptions, have to pay 10–20 SEK (1.09–2.18 EUR, 1.49–2.98 USD) depending on the time of day between 06:30 and 18:29. The maximum tax amount per vehicle per day is 60 SEK (6.53 EUR, 8.94 USD). Payment is done by various means within 14 days after one has passed one of the control points, one cannot pay at the control points.

United Kingdom

Durham became the first city in the UK to have a permanent congestion charge in 2002. London has had a congestion charge in the central area since 2003. Administered by Transport for London (TfL), the charge was initially set at £5, from 17 February 2003, then raised to £8 on 4 July 2005. The daily charge must be paid by the registered keeper of a vehicle that is on public roads in the congestion charge zone between 7 a.m. and 6 p.m. (previously 6:30 p.m.), Monday to Friday. Failure to pay the charge means a fine of at least £50. The charge area was extended into parts of west London on 19 February 2007.

A scheme similar to the one in London was proposed in Manchester, covering a wider area but with a much smaller daily charging window covering the morning and evening rush hours. However, this was overwhelmingly rejected when voted upon in Greater Manchester. A scheme for Cambridge is currently under consideration and the subject of heated public debate, with council surveys showing that a majority of Cambridge-area residents reject the scheme.

A scheme for Edinburgh was rejected in a public referendum in February 2005. On 2008-03-05, councils from across the West Midlands, including those from Birmingham and Coventry, rejected the idea of imposing road pricing schemes on the area, this was despite promises from central government of transport project funding in exchange for the implementation of a road pricing pilot scheme. Similar schemes proposed for cities in the East Midlands have also been dropped.

Extensive studies are being done on introducing a scheme for all UK vehicles, with an aim to implementation at the earliest around 2013. In October 2005 the UK government suggested they explore "piggy-backing" road pricing on private sector technologies, such as usage based insurance (also known as pay-as-you-drive, or PAYD). This method would avoid a large-scale public sector procurement exercise, but such products are unlikely to penetrate the mass market. If introduced, this scheme would likely see a charge being levied per kilometre depending on the time of day, the road being driven along, and perhaps the type of vehicle. For example, a large car driving along the western section of the M25 in rush hour would pay a high charge; a small car driving along a rural lane would pay a much lower charge.

The very highest charges would be likely in the most congested urban areas. It is expected that rural motorists would benefit the most from such a scheme, perhaps by paying less through road pricing than they do at present through petrol and car taxes, whereas urban motorists would pay much more than they presently do. However, this is highly dependent on whether such a scheme would be designed to be either revenue neutral or congestion neutral. A revenue neutral scheme would replace (at least in part) petrol and vehicle taxes, and would be such that Treasury revenue under the new scheme would equal the revenue from current taxes.

A congestion neutral scheme would be designed so that growth in congestion levels would stop as a result of the new charges; the latter scheme would require significantly higher (and increasingly higher) charges than the revenue neutral scheme and so would be unpopular with the UK's 30 million motorists. The carbon emission consequence of moving from fuel duty to a charge per mile has been raised as a concern by some environmentalists, as has any diversionary response from heavily trafficked (and hence more expensive) roads. The UK government announced funding for road pricing research in seven local areas in November 2005.

In June 2005, Transport Secretary Alistair Darling announced the current proposals to introduce road pricing. Every vehicle would be fitted with a satellite receiver to calculate charges, with prices (including fuel duty) ranging from 2p per mile on uncongested roads to £1.34 on the most congested roads at peak times.

A 2007 online petition against road pricing, started by Peter Roberts and hosted by the British government attracted over 1.8 million signatures, equivalent to 6% of the entire driving population.

Over 150,000 signatures were added during the last day before the petition closed on 20 February 2007. In reply, the prime minister e-mailed the petitioners outlining his rationale, denying that the proposals were to introduce a stealth tax or increase surveillance, and promising 'debate' before a decision was made as to whether to introduce a national scheme. Also, in a recent poll 74% of those questioned opposed road pricing.

In July 2008, Roberts started the Drivers' Alliance, an organisation dedicated to researching the issues surrounding road pricing and campaigning against its introduction. The Drivers' Alliance was instrumental in preventing the introduction of a congestion charge in Manchester and also changes in UK government policy where road pricing is no longer being considered.

Toll Road

A toll road (or tollway, turnpike, pike, toll highway or an express toll route) is a privately or publicly built road for which a driver pays a toll (a fee) for use. Structures for which tolls are charged include toll bridges and toll tunnels. Non-toll roads are financed using other sources of revenue, most typically fuel tax or general tax funds. The building or facility in which a toll is collected may be called a toll booth, toll house, toll plaza, toll station, toll bar or toll gate. This building is usually found on either side of a bridge and at exits.

Variations

Three systems of toll roads exist: open (with mainline barrier toll plazas); closed (with entry/exit tolls) and all-electronic toll collection (no toll booths, only electronic toll collection gantries at entrances and exits, or at strategic locations on the mainline of the road).

On an open toll system, all vehicles stop at various locations along the highway to pay a toll. While this may save money from the lack of need to construct tolls at every exit, it can cause traffic congestion, and drivers may be able to avoid tolls (shunpike) by exiting and re-entering the highway.

With a closed system, vehicles collect a ticket when entering the highway. In some cases, the ticket displays the toll to be paid on exit. Upon exit, the driver must pay the amount listed for the given exit. Should the ticket be lost, a driver must typically pay the maximum amount possible for travel on that highway. Short toll roads with no intermediate entries or exits may have only one toll plaza at one end, with motorists travelling in either direction paying a flat fee either when they enter or when they exit the toll road. In a variant of the closed toll system, mainline barriers are present at the two endpoints of the toll road, and each interchange has a ramp toll that is paid upon exit or entry. In this case, a motorist pays a flat fee at the ramp toll and another flat fee at the end of the toll road; no ticket is necessary. In an all-electronic system (such as that used on Highway 407 in the Canadian province of Ontario and the Fort Bend Westpark Tollway in the U.S. state of Texas), no cash toll collection takes place, tolls are usually collected with the use of a transponder mounted on the windshield of each vehicle, which is linked to a customer account which is debited for each use of the toll road. On some roads, such as Highway 407, automobiles and light trucks without transponders are permitted to use the road (though trucks with a gross vehicle weight over 5,000 kilograms must have a transponder)-a bill for the toll due is then sent to the registered owner of the vehicle by mail; by contrast, the Fort Bend Westpark Tollway requires all vehicles to be equipped with a transponder.

Modern toll roads often use a combination of the three, with various entry and exit tolls supplemented by occasional mainline tolls.

Some toll roads charge a toll in only one direction, such as where the M4 in Great Britain crosses the River Severn on either of the two Severn Bridges. On these bridges, it is free to travel from Wales into England, but a toll must be paid on the return journey. This is only practical where the detour to avoid the toll is very large – in this case about 40 miles.

Toll payments may be made in cash, by credit card, by pre-paid card, or by an electronic toll collection system. In some European countries, payment is made using stickers which are affixed to the windscreen. Some toll booths are automated. Tolls may vary according to the distance traveled, the building and maintenance costs of the motorway, and the type of vehicle.

Early Toll Roads

Tolls have been placed on roads at various times in history, often to generate funds for repayment of toll revenue bonds used to finance constructions and/or operation

Toll roads are at least 2700 years old, as tolls had to be paid by travellers using the Susa–Babylon highway under the regime of Ashurbanipal, who reigned in the seventh century BC. Aristotle and Pliny refer to tolls in Arabia and other parts of Asia. In India, before the 4th century BC, the Arthasastra notes the use of tolls. Germanic tribes charged tolls to travellers across mountain passes. Tolls were used in the Holy Roman Empire in the 14th century and 15th century.

A 14th century example (though not for a road) is Castle Loevestein in the Netherlands, which was built at a strategic point where 2 rivers meet, and charged tolls on boats sailing along the river.

Many modern European roads were originally constructed as toll roads in order to recoup the costs of construction. In 14th century England, some of the most heavily used roads were repaired with money raised from tolls by pavage grants. Turnpike trusts were established in England from 1706 onwards, and were ultimately responsible for the maintenance and improvement of most main roads in England and Wales, until they were gradually abolished from the 1870s. Most trusts improved existing roads, but some new ones, usually only short stretches of road, were also built. Thomas Telford's Holyhead road (now the A5 road) is exceptional as a particularly long new road, built in the early 19th century with many toll booths along its length.

Some cities in Canada had toll roads in the 19th Century. Roads radiating from Toronto required users to pay at toll gates along the street (Yonge Street, Bloor Street, Davenport Road, Kingston Road) and disappeared after 1895.

19th century plank roads were usually operated as toll roads. One of the first U.S. motor roads, the Long Island Motor Parkway (which opened on October 10, 1908) was built by William Kissam Vanderbilt II, the great-grandson of Cornelius Vanderbilt. The road was closed in 1938 when it was taken over by the state of New York in lieu of back taxes.

National Toll-road Differences

Toll roads are found in many countries. The way they are funded and operated may differ from country to country. Some of these toll roads are privately owned and operated. Others are owned by the government. Some of the government-owned toll roads are privately operated.

Some toll roads are managed under such systems as the Build-Operate-Transfer (BOT) system. Private companies build the roads and are given a limited franchise. Ownership is transferred to the government when the franchise expires. Throughout the world, this type of arrangement is prevalent in Australia, India, South Korea, Japan, Philippines, and Canada.

The (BOT) system is a fairly new concept that is gaining ground in the United States, with Arkansas, California, Delaware, Florida, Illinois, Indiana, Mississippi, Texas, and Virginia already building and operating toll roads under this scheme. Pennsylvania, Massachusetts, New Jersey, and Tennessee are also considering the BOT methodology for future highway projects.

The more traditional means of managing toll roads in the United States is through semi-autonomous public authorities. New York, Massachusetts, New Hampshire, New Jersey, Maryland, Ohio, Pennsylvania, Kansas, Oklahoma, and West Virginia manage their toll roads in this manner. While most of the toll roads in California, Delaware, Florida, Texas, and Virginia are operating under the BOT arrangement, a few of the older toll roads in these states are still operated by public authorities.

In France, all toll roads are operated by private companies, and the government takes a part of their profit.

Critics of Toll Roads

Toll roads have been criticized as being inefficient in various ways:

1. They require vehicles to stop or slow down, manual toll collection wastes time and raises vehicle operating costs.
2. Collection costs can absorb up to one-third of revenues, and revenue theft is considered to be comparatively easy.
3. Where the tolled roads are less congested than the parallel "free" roads, the traffic diversion resulting from the tolls increases congestion on the road system and reduces its usefulness.
4. By tracking the vehicle locations, their drivers are subject to an effectual restriction of their freedom of movement and freedom from excessive surveillance.

Toll Collection Technology

An adaptation of military "identification friend or foe" or RFID technology, called electronic toll collection, is lessening the delay incurred in toll collection. The electronic system determines whether a passing car is enrolled in the program, alerts enforcers if it is not. The accounts of registered cars are debited automatically without stopping or even opening a window. Currently, DSRC is used as a wireless protocol. Other systems are based on GPRS/GSM and GPS technology. Such a system (for trucks only) in Germany launched successfully in January 2005 and by the end of its first year of operation will have charged tolls for around 22 billion driven kilometres.

One of the advantages of GPS-based systems is their ability to adapt easily and quickly to changes in charge parameters (road classes, vehicle types, emission levels, time slots, etc.). Another advantage is the systems' ability to support other value-added services on the same technology platform. These services might include fleet and vehicle engine management systems, emergency response services, pay-as-you-drive insurance services and navigation capabilities.

The first major deployment of an RFID electronic toll collection system in the United States was on the Dallas North Tollway in 1989 by Amtech. The Amtech RFID technology used on the Dallas North Tollway was originally developed at Sandia Labs for use in tagging and tracking livestock. In the same year, the Telepass active transponder RFID system was introduced across Italy.

Highway 407 in the province of Ontario, Canada has no toll booths, and instead reads a transponder mounted on the windshields of each vehicle using the road (the rear license plates of vehicles lacking a transponder are photographed when they enter and exit the highway). This made the highway the first all-automated highway in the world. A bill is mailed monthly for usage of the 407. Lower charges are levied on frequent 407 users who carry electronic transponders in their vehicles. The approach has not been without controversy: In 2003 the 407 ETR settled PDF a class action with a refund to users. The same method is used on Highway 6 in Israel and the reversible lanes of the Lee Roy Selmon Crosstown Expressway in Hillsborough County, Florida (in the latter case, the system reads Sun Pass transponders).

Throughout most of the East Coast of the United States, E-ZPass (operated under the brands I-Pass in Illinois, i-Zoom in Indiana, and Fast Lane in Massachusetts) is accepted on almost all toll roads. Similar systems include Sun Pass in Florida and Fastrak in California. The systems use a small radio transponder mounted in or on a customer's vehicle to deduct toll fares from a pre-paid account as the vehicle passes through the toll barrier. This reduces manpower at toll booths and increases traffic flow and fuel efficiency by reducing the need for complete stops to pay tolls at these locations.

By designing a tollgate specifically for electronic collection, it is possible to carry out open-road tolling, where the customer does not need to slow at all when passing through the tollgate. The U.S. state of Texas is testing a system on a stretch of Texas 121 that has no toll booths. Drivers without a Toll Tag have their license plate photographed automatically and the registered owner will receive a monthly bill, at a higher rate than those vehicles with Toll Tags.

Another feature of many electronic toll collection systems is inter-agency interoperability, where the same transponder is accepted at many toll agencies. For instance, the E-ZPass tag is accepted at most toll facilities in the Eastern United States, from Virginia to Maine, west to the Peace Bridge spanning the Niagara River, and in Ohio, Indiana, and Illinois. The TxTAG system allows interoperability throughout the state of Texas, but is not compatible with systems used outside of Texas. Electronic toll collection systems also have drawbacks. A computer glitch can result in delays several miles long. Some U.S. state turnpike commissions have debated implementing E-ZPass but have found that such a system would be ineffective because most of the people who use the turnpike are not commuters, are from states that have no ETS on turnpikes, or are from states that don't have a turnpike at all. The toll plazas of some turnpikes are antiquated because they were originally built for traffic that stops to pay the toll or get a ticket.

The technology does have its limits. For instance, the Highway 407 automatic number plate recognition technology has a reputation for the occasional misread plate, leading to bills being sent to motorists in remote parts of Ontario who have never been near the tollway. The Ontario government responded to complaints by hiring an ombudsman to address 407 toll complaints.

Closed System

For toll roads, a "closed system" refers to a road where a motorist obtains a ticket upon entering the toll road, then pays a toll upon exiting the expressway. The toll is calculated by the distance travelled on the toll road. In the United States, for instance, the Kansas Turnpike, Pennsylvania Turnpike, Ohio Turnpike, and portions of Florida's Turnpike currently implement closed systems. In contrast, a toll road using an 'open system' consists of mainline toll plazas (a.k.a., toll barriers) at set intervals; it is possible for motorists to get on an 'open toll road' after one toll barrier and exit before the next one, thus travelling on the toll road toll-free. Most open toll roads have ramp tolls or partial access junctions to prevent this. The Massachusetts Turnpike or "MassPike", implements both systems in different sections.

Asian Application

Singapore

Singapore implemented the world's first congestion pricing scheme in 1975, through manual police control around the CBD of an urban area. In September 1998 the system was upgraded with Etc. technology, 100% free-flow. The electronic toll collection scheme adopted was implemented by the Land Transport Authority (LTA), after careful planning and successfully stress-testing the system. The congestion charges were implemented as part of a comprehensive package of road pricing and harsh ownership restraints, in recognition of the country's land constraints, need of economic competitiveness, and to avoid the traffic gridlock that chokes many cities in the world.

One key aspect of traffic management in Singapore is the restraint of vehicle ownership, either through the imposition of high ownership costs or restriction on the actual growth of the car population. These measures have included high annual road tax, custom duties and vehicle registration fees.

Besides fiscal deterrents, supply of motor vehicles was regulated since 1990, when a Vehicle Quota System was introduced. Use-related charges, such as fuel taxes (50% of final sale price), congestion charges and high parking rates are utilized by public authorities to further constraint travel. In parallel to the whole spectrum of road pricing measures, the government has invested heavily in public transportation and implemented a park-and-ride scheme. In summary, Singapore's urban and transport strategy allowed the users to have pro-transit "carrots" matching auto-restraint "sticks", As a result, and despite having one of the highest per capita incomes in Asia, fewer than 30% of Singaporean households owns cars.

In an effort to improve the pricing mechanism and to introduce real-time variable pricing, Singapore's LTA, together with IBM, ran a pilot from December 2006 to April 2007, with a traffic estimation and prediction tool, which uses historical traffic data and real-time feeds with flow conditions from several sources, in order to predict the levels of congestion up to an hour in advance. By accurate estimating prevailing and emerging traffic conditions, this technology is expected to allow variable pricing, together with improved overall traffic management, including the provision of information in advanced to alert drivers about conditions ahead, and the prices being charged at that moment. This new system integrates with the various LTA's traffic management existing systems, such as the Green Link

Determining System, Traffic Scan, Expressway Monitoring Advisory System, Junction Electronic Eyes, and the Electronic Road Pricing system. The pilot results were successful, showing overall prediction results above 85 percent of accuracy.

Shanghai

Following the strategy of Singapore, the city of Shanghai has implemented policies to restrain both car use and ownership, while improving public transport in Shanghai. Since 1998, the number of new car registrations is limited to 50,000 vehicles a year. Car registrations are sold in a public auction, with prices reaching up to US$5,000 in 2006. Also, parking is limited and there are restrictions on getting a driver's license. Main roadways and highways are tolled, and an assessment was completed to evaluate implementation of congestion pricing for vehicles entering the central business district. The City of Nanjing is also considering the implementation of congestion pricing.

United States

New York Proposal

On April 22, 2007, New York City Mayor Michael Bloomberg, citing what he considered to be successes in London, Singapore and Stockholm, proposed a plan to charge $8 per day for cars to use the streets of the central business district (southern half of Manhattan) but not when using only the marginal highways, or nights or weekends. It would not involve satellite location, but drivers who wanted their tolls collected automatically could have a transponder like the E-ZPass already used to collect tolls on tunnels and bridges.

Immediately following the April 22 announcement, a coalition under the banner Campaign for New York's Future came out in support of the Mayor's sustainability proposal, PlaNYC 2030. Others opposed it, saying it would create "rat run" districts at the border.

> *On July 16, 2007, the New York Legislature shelved the proposal to bring congestion pricing to Manhattan. A week later they passed a law creating a 17-member New York City Traffic Congestion Mitigation Commission to study methods. The Commission's report was favourable and the City Council voted for the measure, but in April 2008 the New York Legislature declined to vote on it, stalling the initiative.*

San Francisco Proposal

In 2006, San Francisco authorities began a feasibility study to evaluate congestion pricing in the city. The study, called the Mobility, Access and Pricing Study (MAPS), was financed with a US$1 million grant from the Federal Highway Administration's Value Pricing Program. The study is part of a congestion pricing demonstration project under the Urban Partnerships Congestion Initiative, for which the San Francisco Bay Area was awarded a $158 million grant.

Initial results from the study show that the program is feasible, and that typical difficulties and controversial issues had been addressed. Authorities are considering exempting low-income drivers and residents within the toll zones. Discounts for commercial fleets were also considered. Different pricing scenarios were analyzed and presented in public meetings in December 2008, and the final study results are expected in 2009.

Criticisms

Opposition to road pricing, when coming from the broad political left is largely directed at perceptions of fairness. Charging for something that was once "free" may be seen as unfair. Road pricing has the possibility of being a regressive tax, in that a flat-rate tax falls more heavily on poor drivers than the rich. A way for the government to deflect this criticism would be to use the toll revenue to reduce other, equally regressive taxes. New toll roads in a largely free system may be seen as punishing one area when the rest of the system enjoys toll-free motoring.

Proponents of pricing would counter the fairness or equity argument by stating that pricing creates a choice, and choices are fair because people are not identical, sometimes people have high values of time (e.g., when they are late for an appointment), sometimes they have lower values of time (e.g., when they are enjoying the drive). The proponents would thus suggest that making all drivers pay the same tax to receive the same service isn't fair if people value the service differently. Another argument is that, while road pricing may be unfair to *some* road users, the alternative, i.e., congestion is unfair to *all* road users, since it wastes everyone's resources. The ultimate fairness of road pricing is only determined once the use of any net revenues is taken into account.

Conservative critics such as Steven Norris, on the other hand, say that "free" roads produce positive externalities that outweigh the opportunity cost of congestion, i.e., that road pricing reduces the overall number of journeys, thus harming business and economic growth. In particular, Steven Norris argues that the cost of the congestion charge disproportionately hits low-paid workers whose working hours start at night when public transport is not available and end when the congestion period is in force, indirectly hitting London's service economy.

Motoring interest groups see road pricing as an additional financial burden on already allegedly over-taxed car owners. Many are not opposed to road tolls as such, but wish to see them as a replacement for fuel tax rather than an additional charge.

Some groups of libertarian inclination, such as the Association of British Drivers, criticize road pricing on the basis of individual rights. They argue that freedom of movement is a fundamental right that should not be infringed through financial barriers, and sometimes compare the practice to highwaymen. Some libertarians, however, generally favour transfer of roads to private ownership, because they believe that, within a free market, competition will force prices down and quality up, and for the moral reason that those who don't use the road will no longer be forced to pay for it via taxes, and that only those who drive on the roads will need to pay for their maintenance.

Others see proposed schemes such as PAYD (which is based upon a compulsory GPS tracking system) as an infringement on their rights to privacy, and fear that such a vast surveillance system may be abused. However, such systems need not necessarily invade the privacy of road users. For example, users of the Express Lanes on California State Route 91 were able to open an account with the California Private Transportation Company without revealing their real name. As long as the account was in credit, no information about the payer needed to enter the computer system. Fewer than five Californians sought to protect their privacy in this way.

The general public appears to have concerns about the proposed introduction of road pricing in the UK, with fears that the government could be using it to increase motoring taxes overall. In 2003, the Institute of Public Policy Research think-tank concluded that overall road pricing would have to raise more money than current taxes if it were to reduce congestion. Local authorities have been denied access to government transport funds for not including proposals for road pricing in their applications.

Congestion Pricing

Congestion pricing is an efficiency pricing strategy that requires the users to pay more for that public good, thus increasing the welfare gain or net benefit for society. Congestion pricing is one of a number of alternative demand side (as opposed to supply side) strategies offered by economists to address congestion. Congestion pricing was first implemented in Singapore in 1975, together with a comprehensive package of road pricing measures, stringent car ownership rules and improvements in mass transit. Thanks to technological advances in electronic toll collection, Singapore upgraded its system in 1998.

Similar pricing schemes were implemented in Rome in 2001, as an upgrade to the manual zone control system implemented in 1998; London in 2003 and extended in 2007; Stockholm in 2006, as seven month trial, and then on a permanent basis since August 2007; and since January 2008, Milan introduced a traffic charge scheme as a one-year trial, called Ecopass, that exempts higher emission standard vehicles (Euro IV) and other alternative fuel vehicles. Later during the year the Ecopass was extended until December 31, 2009.

Even the transport economists who advocate congestion pricing have anticipated several practical limitations, concerns and controversial issues regarding the actual implementation of this policy. As summarized by Cervero: "True social-cost pricing of metropolitan travel has proven to be a theoretical ideal that so far has eluded real-world implementation.

The primary obstacle is that except for professors of transportation economics and a cadre of vocal environmentalists, few people are in favour of considerably higher charges for peak-period travel.

Middle-class motorists often complain they already pay too much in gasoline taxes and registration fees to drive their cars, and that to pay more during congested periods would add insult to injury. In the United States, few politicians are willing to champion the cause of congestion pricing in fear of reprisal from their constituents...

Critics also argue that charging more to drive is elitist policy, pricing the poor off of roads so that the wealthy can move about unencumbered. It is for all these reasons that peak-periord pricing remains a pipe dream in the minds of many."

Road Space Rationing

Transport economists consider road space rationing an alternative to congestion pricing, but road space rationing is considered more equitable, as the restrictions force all drivers to reduce auto travel, while congestion pricing restrains less those who can afford paying the congestion charge. Nevertheless, high-income users can avoid the restrictions by owning a second car. Road space rationing based on license numbers has been implemented in cities such as Athens (1982), Mexico City (1989), Sao Paulo (1997), Santiago, Chile, Bogota, Colombia, La Paz (2003), Bolivia, and San Jose (2005), Costa Rica.

Tradable Mobility Credits

A more acceptable policy on automobile travel restrictions, proposed by transport economists to avoid inequality and revenue allocation issues, is to implement a rationing of peak period travel but through revenue-neutral credit-based congestion pricing. This concept is similar to the existing system of emissions trading of carbon credits, proposed by the Kyoto Protocol to curb greenhouse emissions. Metropolitan area or city residents, or the taxpayers, will have the option to use the local government-issued mobility rights or congestion credits for themselves, or to trade or sell them to anyone willing to continue travelling by automobile beyond the personal quota. This trading system will allow direct benefits to be accrued by those users shifting to public transportation or by those reducing their peak-hour travel rather than the government.

Funding and Financing

Methods of funding and financing transport network maintenance, improvement and expansion are debated extensively and form part of the transport economics field.

Funding issues relate to the ways in which money is raised for the supply of transport capacity. Taxation and user fees are the main methods of fund-raising. Taxation may be general (e.g. income tax), local (e.g. sales tax or land value tax) or variable (e.g. fuel tax), and user fees may be tolls, congestion charges or fares). The method of funding often attracts strong political and public debate.

Financing issues relate to the way in which these funds are used to pay for the supply of transport. Loans, bonds, public-private partnerships and concessions are all methods of financing transport investment.

Regulation & Competition

Regulation of the supply of transport capacity relates to both safety regulation and economic regulation. Transport economics considers issues of the economic regulation of the supply of transport, particularly in relation to whether transport services and networks are provided by the public sector (i.e. socially), by the private sector (i.e. competitively) or using a mixture of both.

Transport networks and services can take on any combination of regulated/deregulated and public/private provision. For example, bus services in the UK outside London are provided by both the public and private sectors in a deregulated economic environment (where no-one specifies which services are to be provided, so the provision of services is influenced by the market), whereas bus services within London are provided by the private sector in a regulated economic environment (where the public sector specifies the services to be provided and the private sector competes for the right to supply those services-i.e. franchising).

The regulation of public transport is often designed to achieve some social, geographic and temporal equity as market forces might otherwise lead to services being limited to the most popular travel times along the most densely settled corridors of development. National, regional or municipal taxes are often deployed to provide a network that is socially acceptable (e.g. extending timetables through the daytime, weekend, holiday or evening periods and intensifying the mesh of routes beyond that which a lightly regulated market would probably provide).

Franchising may be used to create a supply of transport that balances the free-market supply outcome and the most socially desirable supply outcome.

Project Appraisal and Evaluation

The most sophisticated methods of project appraisal and evaluation have been developed and applied in the transport sector. It should be noted that the terms 'appraisal' and 'evaluation' are often confused in relation to the assessment of projects. Appraisal refers to *ex ante* (before the event) assessment and evaluation refers to *ex post* (after the event) assessment.

Appraisal

The appraisal of changes in the transport network is one of the most important applications of transport economics. In order to make an assessment of whether any given transport project should be carried out, transport economics can be used to compare the costs of the project with its benefits (both social and financial). Such an assessment is known as a cost-benefit analysis, and is usually a fundamental piece of information for decision-makers, as it places a value on the net benefits (or disbenefits) of schemes and generates a ratio of benefits to costs which may be used to prioritise projects when funding is constrained.

A primary difficulty in project appraisal is the valuation of time. Travel time savings are often cited as a key benefit of transport projects, but people in different occupations, carrying out different activities and in different social classes value time differently.

Appraising projects on the basis of their supposed reductions in travel times has come under scrutiny in recent years with the recognition that improvements in capacity generate trips that would not have been made (induced demand), partially eroding the benefits of reduced travel times. Therefore an alternative method of appraisal is to measure changes in land value and consumer benefits from a transport project rather than the measuring benefits accruing to travellers themselves. However, this method of analysis is much more difficult to carry out.

Another problem is that many transport projects have impacts that cannot be expressed in monetary terms, such as impacts on, for example, local air quality, biodiversity and community severance. Whilst these impacts can be included in a detailed environmental impact assessment, a key issue has been how to present these assessments alongside estimates of those costs and benefits that can be expressed in monetary terms. Recent developments in transport appraisal practice in some European countries have seen the application of multi-criteria decision analysis based decision support tools. These build on existing cost-benefit analysis and environmental impact assessment techniques and help decision makers weigh up the monetary and non-monetary impacts of transport projects. In the UK, one such application, the New Approach to Appraisal has become a cornerstone of UK transport appraisal.

Evaluation

The evaluation of projects enables decision makers to understand whether the benefits and costs that were estimated in the appraisal materialised. Successful project evaluation requires that the necessary data to carry out the evaluation is specified in advance of carrying out the appraisal.

The appraisal and evaluation of projects form stages within a broader policy making cycle that includes:

- identifying a rationale for a project
- specifying objectives
- appraisal
- monitoring implementation of a project
- evaluation
- feedback to inform future projects.

5

The Modes of Transport and Patterns of Movement

Transport or transportation is the movement of people and goods from one location to another. Modes of transport include air, rail, road, water, cable, pipeline, and space. The field can be divided into infrastructure, vehicles, and operations. Transport infrastructure consists of the fixed installations necessary for transport, and may be roads, railways, airways, waterways, canals and pipelines, and terminals such as airports, railway stations, bus stations, warehouses, trucking terminals, refuelling depots (including fuelling docks and fuel stations), and seaports. Terminals may be used both for interchange of passengers and cargo and for maintenance.

Vehicles travelling on these networks may include automobiles, bicycles, buses, trains, trucks, people, helicopters, and aircraft. Operations deal with the way the vehicles are operated, and the procedures set for this purpose including financing, legalities and policies. In the transport industry, operations and ownership of infrastructure can be either public or private, depending on the country and mode. Passenger transport may be public, where operators provide scheduled services, or private. Freight transport has become focused on containerization, although bulk transport is used for large volumes of durable items.

Transport plays an important part in economic growth and globalization, but most types cause air pollution and use large amounts of land. While it is heavily subsidized by governments, good planning of transport is essential to make traffic flow, and restrain urban sprawl.

Mode

A mode of transport is a solution that makes use of a particular type of vehicle, infrastructure and operation. The transport of a person or of cargo may involve one mode or several modes, with the latter case being called intermodal or multimodal transport. Each mode has its advantages and disadvantages, and will be chosen for a trip on the basis of cost, capability, route, and speed.

Human-powered

Human-powered transport is the transport of people and/or goods using human muscle-power, in the form of walking, running and swimming. Modern technology has allowed machines to enhance human-power. Human-powered transport remains popular for reasons of cost-saving, leisure, physical exercise and environmentalism. Human-powered transport is sometimes the only type available, especially in underdeveloped or inaccessible regions. It is considered an ideal form of sustainable transportation.

Although humans are able to walk without infrastructure, the transport can be enhanced through the use of roads, especially when enforcing the human power with vehicles, such as bicycles and inline skates. Human-powered vehicles have also been developed for difficult environments, such as snow and water, by watercraft rowing and skiing; even the air can be entered with human-powered aircraft.

Animal-powered

Animal-powered transport is the use of working animals for the movement of people and goods. Humans may ride some of the animals directly, use them as pack animals for carrying goods, or harness them, alone or in teams, to pull sleds or wheeled vehicles. Animals are superior to people in their speed, endurance and carrying capacity; prior to the Industrial Revolution they were used for all land transport impracticable for people, and they remain an important mode of transport in less developed areas of the world.

Air

Aviation is the design, development, production, operation, and use of aircraft, especially heavier-than-air aircraft.

History

Many cultures have built devices that travel through the air, from the earliest projectiles such as stones and spears., the boomerang in Australia, the hot air Kongming lantern, and kites. There are early legends of human flight such as the story of Icarus, and Jamshid in Persian myth, and later, somewhat more credible claims of short-distance human flights appear, such as the flying automaton of Archytas of Tarentum (428–347 BC), the winged flights of Abbas Ibn Firnas (810–887), Eilmer of Malmesbury (11th century), and the hot-air Passarola of Bartolomeu Lourenço de Gusmao (1685–1724).

The modern age of aviation began with the first untethered human lighter-than-air flight on November 21, 1783, in a hot air balloon designed by the Montgolfier brothers. The practicality of balloons was limited because they could only travel downwind. It was immediately recognized that a steerable, or dirigible, balloon was required. Jean-Pierre Blanchard flew the first human-powered dirigible in 1784 and crossed the English Channel in one in 1785.

In 1799 Sir George Cayley set forth the concept of the modern airplane as a fixed-wing flying machine with separate systems for lift, propulsion, and control. Early dirigible developments included machine-powered propulsion (Henri Giffard, 1852), rigid frames (David Schwarz, 1896), and improved speed and maneuverability (Alberto Santos-Dumont, 1901)

While there are many competing claims for the earliest powered, heavier-than-air flight, the most widely-accepted date is December 17, 1903 by the Wright brothers. The Wright brothers were the first to fly in a powered and controlled aircraft. Previous flights were gliders (control but no power) or free flight (power but no control), but the Wright brothers combined both, setting the new standard in aviation records. Following this, the widespread adoption of ailerons versus wing warping made aircraft much easier to control, and only a decade later, at the start of World War I, heavier-than-air powered aircraft had become practical for reconnaissance, artillery spotting, and even attacks against ground positions.

Aircraft began to transport people and cargo as designs grew larger and more reliable. In contrast to small non-rigid blimps, giant rigid airships became the first aircraft to transport passengers and cargo over great distances. The best known aircraft of this type were manufactured by the German Zeppelin company.

The most successful Zeppelin was the Graf Zeppelin. It flew over one million miles, including an around-the-world flight in August 1929. However, the dominance of the Zeppelins over the airplanes of that period, which had a range of only a few hundred miles, was diminishing as airplane design advanced. The "Golden Age" of the airships ended on May 6, 1937 when the Hindenburg caught fire, killing 36 people. Although there have been periodic initiatives to revive their use, airships have seen only niche application since that time.

Great progress was made in the field of aviation during the 1920s and 1930s, such as Charles Lindbergh's solo transatlantic flight in 1927, and Charles Kingsford Smith's transpacific flight the following year. One of the most successful designs of this period was the Douglas DC-3, which became the first airliner that was profitable carrying passengers exclusively, starting the modern era of passenger airline service. By the beginning of World War II, many towns and cities had built airports, and there were numerous qualified pilots available. The war brought many innovations to aviation, including the first jet aircraft and the first liquid-fueled rockets.

After WW II, especially in North America, there was a boom in general aviation, both private and commercial, as thousands of pilots were released from military service and many inexpensive war-surplus transport and training aircraft became available. Manufacturers such as Cessna, Piper, and Beechcraft expanded production to provide light aircraft for the new middle-class market.

By the 1950s, the development of civil jets grew, beginning with the de Havilland Comet, though the first widely-used passenger jet was the Boeing 707, because it was much more economical than other planes at the time. At the same time, turboprop propulsion began to appear for smaller commuter planes, making it possible to serve small-volume routes in a much wider range of weather conditions.

Since the 1960s, composite airframes and quieter, more efficient engines have become available, and Concorde provided supersonic passenger service for more than two decades, but the most important lasting innovations have taken place in instrumentation and control. The arrival of solid-state electronics, the Global Positioning System, satellite communications, and increasingly small and powerful computers and LED displays, have dramatically changed the cockpits of airliners and, increasingly, of smaller aircraft as well. Pilots can navigate much more accurately and view terrain, obstructions, and other nearby aircraft on a map or through synthetic vision, even at night or in low visibility.

On June 21, 2004, Space Ship One became the first privately funded aircraft to make a spaceflight, opening the possibility of an aviation market capable of leaving the Earth's atmosphere. Meanwhile, flying prototypes of aircraft powered by alternative fuels, such as ethanol, electricity, and even solar energy, are becoming more common and may soon enter the mainstream, at least for light aircraft.

Civil Aviation

Civil aviation is one of two major categories of flying, representing all non-military aviation, both private and commercial. Most of the countries in the world are members of the International Civil Aviation Organization (ICAO) and work together to establish common standards and recommended practices for civil aviation through that agency.

Civil aviation includes two major categories:

- Scheduled air transport, including all passenger and cargo flights operating on regularly-scheduled routes; and
- General aviation (GA), including all other civil flights, private or commercial.

Although scheduled air transport is the larger operation in terms of passenger numbers, GA is larger in the number of flights (and flight hours, in the U.S.) In the U.S., GA carries 166 million passengers each year, more than any individual airline, though less than all the airlines combined.

Some countries also make a regulatory distinction based on whether aircraft are flown for hire:

- Commercial aviation includes most or all flying done for hire, particularly scheduled service on airlines; and
- Private aviation includes pilots flying for their own purposes (recreation, business meetings, etc.) without receiving any kind of remuneration.

All scheduled air transport is commercial, but general aviation can be either commercial or private. Normally, the pilot, aircraft, and operator must all be authorized to perform commercial operations through separate commercial licensing, registration, and operation certificates.

Civil Aviation Authorities

The Convention on International Civil Aviation (the *Chicago Convention*) was originally established in 1944: it states that signatories should collectively work to harmonize and standardize the use of airspace for safety, efficiency and regularity of air transport. All the States signatory to the Chicago Convention, now 188, are obliged to implement the Standards and Recommended Practices (SARPs) of the

Convention. Each signatory country has a Civil Aviation Authority (CAA) (such as the FAA in the United States) to oversee the following areas of civil aviation:

- Personnel Licensing — regulating the basic training and issuance of licenses and certificates.
- Flight Operations — carrying out safety oversight of commercial operators.
- Airworthiness — issuing certificates of registration and certificates of airworthiness to civil aircraft, and overseeing the safety of maintenance organizations.
- Aerodromes — designing and constructing aerodrome facilities.
- Air Traffic Services — managing the traffic inside of a country's airspace.

Air Transport

There are five major manufacturers of civil transport aircraft (in alphabetical order):

- Airbus, based in Europe
- Boeing, based in the United States
- Bombardier, based in Canada
- Embraer, based in Brazil
- United Aircraft Corporation, based in Russia.

Boeing, Airbus, Ilyushin and Tupolev concentrate on wide-body and narrow-body jet airliners, while Bombardier, Embraer and Sukhoi concentrate on regional airliners. Large networks of specialized parts suppliers from around the world support these manufacturers, who sometimes provide only the initial design and final assembly in their own plants. The Chinese ACAC consortium will also soon enter the civil transport market with its ACAC ARJ21 regional jet.

Until the 1970s, most major airlines were flag carriers, sponsored by their governments and heavily protected from competition. Since then, open skies agreements have resulted in increased competition and choicc for consumers, coupled with falling prices for airlines. The combination of high fuel prices, low fares, high salaries, and crises such as the September 11, 2001 attacks and the SARS epidemic have driven many older airlines to government-bailouts, bankruptcy or mergers. At the same time, low-cost carriers such as Ryanair, Southwest and Westjet have flourished.

General Aviation

General aviation (GA) is one of the two categories of civil aviation. It refers to all flights *other than* military and scheduled airline and regular cargo flights, both private and commercial. General aviation flights range from gliders and powered parachutes to large, non-scheduled cargo jet flights. The majority of the world's air traffic falls into this category, and most of the world's airports serve general aviation exclusively.

General aviation is particularly popular in North America, with over 6,300 airports available for public use by pilots of general aviation aircraft (around 5,300 airports in the U.S., and over 1,000 in Canada). In comparison, scheduled flights operate from around 600 airports in the U.S. According to the U.S. Aircraft Owners and Pilots Association, general aviation provides more than one percent of the United States' GDP, accounting for 1.3 million jobs in professional services and manufacturing.

General aviation covers a large range of activities, both commercial and non-commercial, including private flying, flight training, air ambulance, police aircraft, aerial firefighting, air charter, bush flying, gliding, and many others. Experimental aircraft, light-sport aircraft and very light jets have emerged in recent years as new trends in general aviation.

Regulation and Safety

Most countries have authorities that oversee all civil aviation, including general aviation, adhering to the standardized codes of the International Civil Aviation Organization (ICAO). Examples include the Federal Aviation Administration (FAA) in the United States, the Civil Aviation Authority (CAA) in Great Britain, the Luftfahrt-Bundesamt (LBA) in Germany, and Transport Canada in Canada.

Since it includes both non-scheduled commercial operations and private operations, with aircraft of many different types and sizes, and pilots with a variety of different training and experience levels, it is not possible to make blanket statements about the regulation or safety record of general aviation. At one extreme, in most countries business jets and large cargo jets face most of the same regulations as scheduled air transport and fly mostly to the same airports. Commercial bush flying and air ambulance operations normally do not operate under as heavy a regulatory burden, and often only use small airports or off-airport strips, where there is less governmental oversight. Aviation accident rate statistics are necessarily estimates. According to the U.S. National Transportation Safety Board, in 2005 general aviation in the United States (excluding charter) suffered 1.31 fatal accidents for every 100,000 hours of flying in that country, compared to 0.016 for scheduled airline flights. In Canada, recreational flying accounted for 0.7 fatal accidents for every 1000 aircraft, while air taxi accounted for 1.1 fatal accident for every 100,000 hours.

Military Aviation

Simple balloons were used as surveillance aircraft as early as the 18th century. Over the years, military aircraft have been built to meet ever increasing capability requirements. Manufacturers of military aircraft compete for contracts to supply their government's arsenal. Aircraft are selected based on factors like cost, performance, and the speed of production.

Types of military aviation:

- Fighter aircraft's primary function is to destroy other aircraft. (e.g. Sopwith Camel, A6M Zero, F-15, MiG-29, Su-27, F-22).
- Ground attack aircraft are used against tactical earth-bound targets. (e.g. Junkers Stuka dive bomber, A-10 Warthog, Ilyushin Il-2, J-22 Orao, and Sukhoi Su-25).
- Bombers are generally used against more strategic targets, such as factories and oil fields. (e.g. Zeppelin, B-29 Superfortress, Tu-95, Dassault Mirage IV, and the B-52 Stratofortress)
- Cargo transport aircraft are used to transport hardware and personnel, such as the C-17 Globemaster III or C-130 Hercules.
- Projectile is used for goods only, normally explosives, but also things like leaflets
- Surveillance aircraft are used for reconnaissance (e.g. Rumpler Taube, de Havilland Mosquito, U-2, and MiG-25R).
- Helicopters are used for assault support, cargo transport and close air support (e.g.AH-64, Mi-24).

Air Traffic Control (ATC)

Air traffic control (ATC) involves communication with aircraft to help maintain *separation* — that is, they ensure that aircraft are sufficiently far enough apart horizontally or vertically for no risk of collision. Controllers may co-ordinate position reports provided by pilots, or in high traffic areas (such as the United States) they may use radar to see aircraft positions.

There are generally four different types of ATC:

- centre controllers, who control aircraft en route between airports
- control towers (including tower, ground control, clearance delivery, and other services), which control aircraft within a small distance (typically 10–15 km horizontal, and 1,000 m vertical) of an airport.
- oceanic controllers, who control aircraft over international waters between continents, generally without radar service.
- terminal controllers, who control aircraft in a wider area (typically 50–80 km) around busy airports.

ATC is especially important for aircraft flying under Instrument flight rules (IFR), where they may be in weather conditions that do not allow the pilots to see other aircraft. However, in very high-traffic areas, especially near major airports, aircraft flying under Visual flight rules (VFR) are also required to follow instructions from ATC.

In addition to separation from other aircraft, ATC may provide weather advisors, terrain separation, navigation assistance, and other services to pilots, depending on their workload.

ATC do not control all flights. The majority of VFR flights in North America are not required to talk to ATC (unless they are passing through a busy terminal area or using a major airport), and in many areas, such as northern Canada and low altitude in northern Scotland, ATC services are not available even for IFR flights at lower altitudes.

Environmental Impact

Like all activities involving combustion, operating powered aircraft (from airliners to hot air balloons) release soot and other pollutants into the atmosphere. Greenhouse gases such as carbon dioxide (CO_2) are also produced. In addition, there are environmental impacts specific to aviation:

- Aircraft operating at high altitudes near the tropopause (mainly large jet airliners) emit aerosols and leave contrails, both of which can increase cirrus cloud formation — cloud cover may have increased by up to 0.2% since the birth of aviation.
- Aircraft operating at high altitudes near the tropopause can also release chemicals that interact with greenhouse gases at those altitudes, particularly nitrogen compounds, which interact with ozone, increasing ozone concentrations.

- Most light piston aircraft burn avgas, which contains tetraethyl lead (TEL), a highly-toxic substance that can cause soil contamination at airports. Some lower-compression piston engines can operate on unleaded mogas, and turbine engines and diesel engines — neither of which requires lead — are appearing on some newer light aircraft.

A fixed-wing aircraft, commonly called airplane, is a heavier-than-air craft where movement of the air in relation to the wings is used to generate lift. The term is used to distinguish from rotary-wing aircraft, where the movement of the lift surfaces relative to the air generates lift. A gyroplane is both fixed-wing and rotary-wing. Fixed-wing aircraft range from small trainers and recreational aircraft to large airliners and military cargo aircraft.

Two things necessary for aircraft are air flow over the wings for lift and an area for landing. The majority of aircraft also need an airport with the infrastructure to receive maintenance, restocking, refuelling and for the loading and unloading of crew, cargo and passengers. While the vast majority of aircraft land and take off on land, some are capable of take off and landing on ice, snow and calm water.

The aircraft is the second fastest method of transport, after the rocket. Commercial jets can reach up to 875 kilometres per hour (544 mph), single-engine aircraft 175 kilometres per hour (109 mph). Aviation is able to quickly transport people and limited amounts of cargo over longer distances, but incur high costs and energy use; for short distances or in inaccessible places helicopters can be used. WHO estimates that up to 500,000 people are on planes at any time.

Rail

Rail Transport

Rail transport is the means of conveyance of passengers and goods by way of wheeled vehicles running on rail tracks. In contrast to road transport, where vehicles merely run on a prepared surface, rail vehicles are also directionally guided by the tracks they run on. Track usually consists of steel rails installed on sleepers/ties and ballast, on which the rolling stock, usually fitted with metal wheels, moves. However, other variations are also possible, such as slab track where the rails are fastened to a concrete foundation resting on a prepared subsurface.

Rolling stock in railway transportation systems generally has lower frictional resistance when compared with highway vehicles, and the carriages and wagons can be coupled into longer trains. The operation is carried out by a Railway company, providing transport between train stations or freight customer facilities. Power is provided by locomotives which either draw electrical power from a railway electrification system or produce their own power, usually by diesel engines. Most tracks are accompanied by a signalling system. Railways are a safe land transportation systems when compared to other forms of transportation. Railway transportation is capable of high levels of passenger and cargo utilization and energy efficiency, but is often less flexible and more capital-intensive than highway transportation is, when lower traffic levels are considered.

The oldest, man-hauled railways date to the 6th century B.C. With the English development of the steam engine, it was possible to construct mainline railways, that were a key component of the industrial revolution. Also, railways reduced the costs of shipping, and allowed for fewer lost goods. The change from canals to railways allowed for "national markets" in which prices varied very little from city to city. Studies have shown that the development of the railway was one of the most important technological inventions of the late 19th century in the United States, without which, GDP would have been lower by 7.0% in 1890. In the 1880s, electrified trains were introduced, and also the first tramways and rapid transit systems came into being. During the 1940s and 1950s, the non-electrified railways in most countries had their steam locomotives replaced by diesel-electric locomotives. During the 1960s, electrified high-speed railway systems were introduced in Japan and a few other countries. Other forms of guided ground transportation outside the traditional railway definitions, such as monorail or maglev, have been tried but have seen limited use.

History, Pre-steam

The earliest evidence of a railway was a 6-kilometre (3.7 mi) Diolkos wagonway, which transported boats across the Corinth isthmus in Greece during the 9th century BC. Trucks pushed by slaves ran in grooves in limestone, which provided the track element. The Diolkos ran for over 1300 years.

Railways began reappearing in Europe after the Dark Ages. The earliest known record of a railway in Europe from this period is a stained-glass window in the Minster of Freiburg im Breisgau in

Germany, dating from around 1350. In 1515, Cardinal Matthaus Lang wrote a description of the Reisszug, a funicular railway at the Hohensalzburg Castle in Austria. The line originally used wooden rails and a hemp haulage rope, and was operated by human or animal power. The line still exists, albeit in updated form, and is probably the oldest railway still to operate.

By 1550, narrow gauge railways with wooden rails were common in mines in Europe. By the 17th century, wooden wagonways were common in the United Kingdom for transporting coal from mines to canal wharfs for transshipment to boats. The world's oldest continually working railway, built in 1758, is the Middleton Railway in Leeds. In 1764, the first gravity railroad in the United States was built in Lewiston, New York. The first permanent was the 1810 Leiper Railroad.

The first iron plate rail way made with cast iron plats on top of wooden rails, was taken into use in 1768. This allowed a variation of gauge to be used. At first only balloon loops could be used for turning, but later, movable points were taken into use, that allowed for switching. From the 1790s, iron edge rails began to appear in the United Kingdom. In 1803, William Jessop opened the Surrey Iron Railway in south London, arguably the world's first horse-drawn public railway. Hot rolling iron allowed the brittle, and often uneven, cast iron rails to be replaced by wrought iron in 1805. These were succeeded by steel in 1857.

Age of Steam

The development of the steam engine spurred ideas for mobile steam locomotives that could haul trains on tracks. The first was patented by James Watt in 1794. In 1804, Richard Trevithick demonstrated the first locomotive-hauled train in Merthyr Tydfil, United Kingdom. Accompanied with Andrew Vivian, it ran with mixed success, breaking some of the brittle cast-iron plates. Two years later, the first passenger horse-drawn railway was opened nearby between Swansea and Mumbles. In 1811, John Blenkinsop designed the first successful and practical railway locomotive—a rack railway worked by a steam locomotive between Middleton Colliery and Leeds on the Middleton Railway.

The locomotive, *The Salamanca,* was built the following year. In 1825 George Stephenson built the *Locomotion* for the Stockton and Darlington Railway, north east England, which was the first public steam railway in the world. In 1829 he built *The Rocket* which was entered in and won the Rainhill Trials. This success led to Stephenson establishing his company as the pre-eminent builder of steam locomotives used on railways in the United Kingdom, the United States and much of Europe. In 1830, the first intercity railway, the Liverpool and Manchester Railway, opened.

The gauge was that used for the early wagonways, and had been adopted for the Stockton and Darlington Railway. The 1,435 mm (4 ft $8^1/_2$ in) width became known as the international standard gauge, used by about 60% of the world's railways. This spurred the spread of rail transport outside the UK. The Baltimore and Ohio that opened in 1830 was the first to evolve from a single line to a network in the United States. By 1831, a steam railway connected Albany and Schenectady, New York, a distance of 16 miles, which was covered in 40 minutes. In 1867, the first elevated railway was built in New York. The symbolically important first transcontinental railway was completed in 1869.

Electrification and Dieselisation

Experiments with electrical railways were started by Robert Davidson in 1838. He completed a battery-powered carriage capable of 6.4 km/h (4 mph). The Giant's Causeway Tramway was the first to use electricity fed to the trains en-route, using a third rail, when it opened in 1883. Overhead wires were taken into use in 1888. At first this was taken into use on tramways, that until then had been horse-hauled tramcars. The first conventional electrified railway was the Roslag Line in Sweden. During the 1890s, many large cities, such as London, Paris and New York used the new technology to build rapid transit for urban commuting. In smaller cities, tramways became common, and were often the only mode of public transport until the introduction of buses in the 1920s. In North America, interurbans became a common mode to reach suburban areas. At first all electric railways used direct current, but in 1904, the Spubeital Line in Austria opened with alternating current.

Steam locomotives require large pools of labour to clean, load, maintain and run. After World War II, dramatically increased labour costs in developed countries made steam an increasingly costly form of motive power. At the same time, the war had forced improvements in internal combustion engine technology that made diesel locomotives cheaper and more powerful. This caused many railway companies to initiate programs to convert all unelectrified sections from steam to diesel locomotion.

Following the large-scale construction of motorways after the war, rail transport became less popular for commuting, and air transport started taking large market shares from long-haul passenger trains. Most tramways were either replaced by rapid transits or buses, while high transshipment costs caused short-haul freight trains to become uncompetitive. The 1973 oil crisis led to a change of mind set, and most tram systems that had survived into the 1970s remain today. At the same time, containerization allowed freight trains to become more competitive and participate in intermodal freight transport. With the 1962 introduction of the Shinkansen high-speed rail in Japan, trains could again have a dominant position on intercity travel. During the 1970s, the introduction of automated rapid transit systems allowed cheaper operation. The 1990s saw an increased focus on accessibility and low-floor trains. Many tramways have been upgraded to light rail, and many cities that closed their old tramways have reopened new light railway systems.

Trains

A train is a connected series of rail vehicles that move along the track. Propulsion for the train is provided by a separate locomotive, or from individual motors in self-propelled multiple units. Most trains carry a revenue load, although non-revenue cars exist for the railway's own use, such as for maintenance-of-way purposes. The railroad engineer or engine driver controls the locomotive or other power cars, although people movers and some rapid transits are driverless.

Haulage

Traditionally, trains are pulled using a locomotive. This involved a single or multiple powered vehicles being located at the front of the train, and providing sufficient adhesion to haul the weight of the full train. This remains dominant for freight trains, and is often used for passenger trains.

A push-pull train has the end passenger car equipped with a driver's cab so the engineer can remote-control the locomotive. This allows one of the locomotive hauled trains drawbacks to be removed, since the locomotive need not be moved to the end of the train each time the train changes direction. A railroad car is a vehicle used for the haulage of either passengers or freight.

A multiple unit has powered wheels throughout the whole train. This the used for rapid transit and tram systems, as well as many both short-and long-haul passenger trains. A railcar is a single, self-powered car. Multiple units have a driver's cab at each end of the unit, and were developed following the ability to build electric motors and engines small enough to build under the coach. There are only a few freight multiple units, most of which are high-speed post trains.

Motive Power

Steam locomotives are locomotives with a steam engine that provides adhesion. Coal, petroleum, or wood is burned in a firebox. The heat warms up water in the fire-tube boiler to create pressurized steam. The steam travels through the smokebox before leaving via the chimney. In the process it powers a piston, that transmits power directly through a connecting rod (US: main rod) and a crankpin (US: wristpin) on the driving wheel (US main driver) or to a crank on a driving axle. Steam locomotives have been phased out in most parts of the world for economical and safety reasons.

Electric locomotives draw power from a stationary source via overhead wire or a third rail. Some also or instead use a battery. A transformer in the locomotive converts the high voltage, low current power to low voltage, high current used in the electric motors that power the wheels. Modern locomotives use three-phase AC induction motors. Electric locomotives are the most powerful traction. They are also the cheapest to run and provide less noise and no local air pollution. However, they require high capital investments both for the catenary and the supporting infrastructure. Accordingly, electric traction is used on urban systems, lines with high traffic and for high-speed rail.

Diesel locomotives use a diesel engine as the prime mover. The energy transmission may be either diesel-electric, diesel-mechanical or diesel-hydraulic, but diesel-electric is dominant. Electro-diesel locomotives are built to run as diesel-electric on unelectrified sections, and as an electric locomotive on electrified sections.

Alternative methods of motive power include magnetic levitation, horse-drawn, cable, gravity, pneumatics and gas turbine.

Passenger Trains

A passenger train travels between stations where passengers may embark and disembark. The oversight of the train is the duty of a conductor. Passenger trains are part of public transport, and often make up the stem of the service, with buses feeding to stations.

Intercity trains are long-haul trains that operate with few stops between cities. Trains typically have amenities such as a dining car.

Some lines also provide over-night services with sleeping cars. Some long-haul trains been given a specific name. Regional trains are medium distance trains that connect cities with outlying, surrounding areas, or provide a regional service. Trains make more stops and have lower speeds. Commuter trains serve suburbs of urban areas, providing a daily commuting service. Airport rail links provide quick access from city centres to airports.

Rapid transit is built in large cities and has the highest capacity of any passenger transport system. It is grade separated and commonly built underground or elevated. At street level, smaller trams can be used. Light rails are upgraded trams, that have step-free access, their own right-of-way and sometimes sections underground. Monorail systems operate as elevated, medium capacity systems. A people mover is a driverless, grade-separated train that serves only a few stations, of as a shuttle.

High-speed rail operate at much higher speeds than conventional railways, the limit being regarded at 200 to 320 km/h. High-speed trains are used mostly for long-haul service, and most systems are in Western Europe and East Asia. The speed record is 574.8 km/h (357.2 mph), set by a modified French TGV. Magnetic levitation trains such as the Shanghai airport train use under-riding magnets which attract themselves upward towards the underside of a guideway, and this line has achieved somewhat higher peak speeds in day-to-day operation than conventional high-speed railways, although only over short distances.

Freight Train

A freight train hauls cargo using freight cars specialized for the type of goods. Freight trains can be very efficient, with economy of scale and high energy efficiency. However, their use is reduced by lack of flexibility, often by the need of transshipment at both ends of the trip due to lack of tracks to the points of pick-up and delivery. Authorities often encourage the use of cargo rail transport due to its environmental profile.

Container trains have become the dominant type in the US for non-bulk haulage. Containers can easily be transshipped to other modes, such as ships and trucks, using cranes. This has succeeded the boxcar (wagon-load), where the cargo had to be loaded and unloaded into the train manually. In Europe the sliding wall wagon has largely superseded the ordinary covered wagons. Other types of cars include refrigerator cars, stock cars for livestock and autoracks for road vehicles. When rail is combined with road transport, a roadrailer will allow semi-trailers to be driven onto the train, allowing for easy transition between road and rail.

Bulk handling represents a key advantage for rail transport. Low or even zero transshipment costs combined with energy efficiency and low inventory costs allow trains to handle bulk much cheaper than by road. Typical bulk cargo includes coal, ore, grains and liquids. Bulk is transported in open-topped cars and tank cars.

Infrastructure

Right of Way

Railway tracks are laid upon land owned or leased by the railway company. Owing to the desirability of maintaining modest grades, rails will often be laid in circuitous routes in hilly or mountainous terrain. Route length and grade requirements can be reduced by the use of alternating cuttings, bridges and tunnels—all of which can greatly increase the capital expenditures required to develop a right of way, while significantly reducing operating costs and allowing higher speeds on longer radius curves. In densely urbanized areas, railways are sometimes laid in tunnels to minimize the effects on existing properties.

Trackage

Track consists of two parallel steel rails, anchored perpendicular to members called ties (sleepers) of timber, concrete, steel, or plastic to maintain a consistent distance apart, or gauge. The track guides the conical, flanged wheels, keeping the cars on the track without active steering and therefore allowing trains to be much longer than road vehicles. The rails and ties are usually placed on a foundation made of compressed earth on top of which is placed a bed of ballast to distribute the load from the ties and to prevent the track from buckling as the ground settles over time under the weight of the vehicles passing above.

The ballast also serves as a means of drainage. Some more modern track in special areas is attached by direct fixation without ballast. Track may be prefabricated or assembled in place. By welding rails together to form lengths of continuous welded rail, additional wear and tear on rolling stock caused by the small surface gap at the joints between rails can be counteracted; this also makes for a quieter ride (passenger trains). On curves the outer rail may be at a higher level than the inner rail. This is called superelevation or cant. This reduces the forces tending to displace the track and makes for a more comfortable ride for standing livestock and standing or seated passengers. A given amount of superelevation will be the most effective over a limited range of speeds.

Turnouts, also known as points and switches, are the means of directing a train onto a diverging section of track. Laid similar to normal track, a point typically consists of a frog (common crossing), check rails and two switch rails. The switch rails may be moved left or right, under the control of the signalling system, to determine which path the train will follow.

Spikes in wooden ties can loosen over time, but split and rotten ties may be individually replaced with new wooden ties or concrete substitutes.

Concrete ties can also develop cracks or splits, and can also be replaced individually. Should the rails settle due to soil subsidence, they can be lifted by specialized machinery and additional ballast tamped under the ties to level the rails. Periodically, ballast must be removed and replaced with clean ballast to ensure adequate drainage. Culverts and other passages for water must be kept clear lest water is impounded by the trackbed, causing landslips. Where trackbeds are placed along rivers, additional protection is usually placed to prevent streambank erosion during times of high water. Bridges require inspection and maintenance, since they are subject to large surges of stress in a short period of time when a heavy train crosses.

Signalling

Railway signalling is a system used to control railway traffic safely to prevent trains from colliding. Being guided by fixed rails with low friction, trains are uniquely susceptible to collision since they frequently operate at speeds that do not enable them to stop quickly or within the driver's sighting distance. Most forms of train control involve movement authority being passed from those responsible for each section of a rail network to the train crew. Not all methods require the use of signals, and some systems are specific to single track railways.

The signalling process is traditionally carried out in a signal box, a small building that houses the lever frame required for the signalman to operate switches and signal equipment. These are placed at various intervals along the route of a railway, controlling specified sections of track. More recent technological developments have made such operational doctrine superfluous, with the centralization of signalling operations to regional control rooms. This has been facilitated by the increased use of computers, allowing vast sections of track to be monitored from a single location.

The common method of block signalling divides the track into zones guarded by combinations of block signals, operating rules, and automatic-control devices so that only one train may be in a block at any time.

Electrification

The electrification system provides electrical energy to the trains, so they can operate without a prime mover onboard. This allows lower operating costs, but requires large capital investments along the lines. Mainline and tram systems normally have overhead wires, which hang from poles along the line. Grade-separated rapid transit sometimes use a ground third rail. Power may be fed as direct or alternating current. The most common currencies are 600 and 750 V for tram and rapid transit systems, and 1,500 and 3,000 V for mainlines. The two dominant AC systems are 15 kV AC and 25 kV AC.

Stations

A railway station serves as an area where passengers can board and alight from trains. A goods station is a yard which is exclusively used for loading and unloading cargo. Large passenger stations have at least one building providing conveniences for passengers, such as purchasing tickets and food. Smaller stations typically only consist of a platform. Early stations were sometimes built with both passenger and goods facilities. Platforms are used to allow easy access to the trains, and are connected to each other via underpasses, footbridge and level crossings. Some large stations are built as cul-de-sac, with trains only operating out from one direction. Smaller stations normally serve local residential areas, and may have connection to feeder bus services. Large stations, in particular central stations, serve as the main public transport hub for the city, and have transfer available between rail services, and to rapid transit, tram or bus services.

Operations

A railway has two major components: the rolling stock (the locomotives, passenger coaches, freight cars, etc.) and the infrastructure (the permanent way, tracks, stations, freight facilities, viaducts, tunnels, etc.).

The operation of the railway is through a system of control, originally by mechanical means, but nowadays more usually electronic and computerized.

Intrinsic Factors

Signalling

Signalling systems used to control the movement of traffic may be either of fixed block or moving block variety.

Fixed Block Signalling

Most blocks are 'fixed' blocks, i.e. they delineate a section of track between two defined points. On timetable, train order, and token-based systems, blocks usually start and end at selected stations. On signalling-based systems, blocks usually start and end at signals. Alternatively, cab signalling may be in use. The lengths of blocks are designed to allow trains to operate as frequently as necessary. A lightly-used branch line might have blocks many kilometres long, whilst a busy commuter railway might have blocks a few hundred metres long.

Moving Block Signalling

A disadvantage of fixed blocks is that the faster trains are permitted to run, the longer the stopping distance, and therefore the longer the blocks need to be. This decreases a line's capacity. With moving block, computers are used to calculate a 'safe zone', behind each moving train, which no other train may enter. The system depends on precise knowledge of where each train is and how fast it is moving.

With moving block, lineside signals are not provided, and instructions are passed direct to the trains. It has the advantage of increasing track capacity by allowing trains to run much closer together.

Types of Rail System

Most rail systems serve a number of functions on the same track, carrying local, long distance and commuter passenger trains, and freight trains. The emphasis on each varies by country. Some urban rail transit, rapid transit and light rail systems are isolated from the national system in the cities they serve. Some freight lines serving mines are also isolated, and these are usually owned by the mine company. An industrial railway is a specialized rail system used inside factories or mines. Mountain railways are usually isolated, with special safety systems.

Permanent Way and Railroad Construction

The permanent way trails through the physical geography. The tracks' geometry is limited by the physical geography.

Types of Vehicle

Trains are pushed/pulled by one or more locomotive units. Two or more locomotives coupled in multiple traction are frequently used in freight trains. Railroad cars or rolling stock consist of passenger cars, freight cars, maintenance cars and in America cabooses. Modern passenger trains sometimes are pushed/pulled by a tail and head unit, of which not both need to be motorised or running. Many passenger trains consist of multiple units with motors mounted beneath the passenger cars.

Passenger Operations

Most public transport passenger operations happen in the train station and in the passenger car. The passenger buys a ticket, either in the station, or on the train (sometimes at a higher fare). There are two ways of validating a ticket. In one case the passenger validates the ticket himself (by perforating it, for instance) and this is randomly checked by a ticket controller. A conductor checks all persons on the train, validates the ticket and devaluates it, so it cannot be used again. Some passenger cars, especially in long distance high speed trains have a restaurant or bar. These need to be catered. In recent times, train catering has been diminished somewhat by vending machines in the train station or on the train.

Freight Operations

Freight or cargo trains are loaded and unloaded in intermodal terminals (also called container freight stations or freight terminals), and at customer locations (e.g. mines, grain elevators, factories).

Intermodal freight transport uses standardized containers, which are handled by cranes. Along their routes, freight trains are routed through rail yards to sort cars and assemble trains for their final destinations, as well as for equipment maintenance, refuelling, and crew changes. Within a freight yard, trains are composed in a classification yard. Switcher or shunter locomotives help the composing.

A unit train (also called a block train), which carries a block of cars all of the same origin and destination, does not get sorted in a classification yard, but may stop in a freight yard for inspection, engine servicing and/or crew changes.

Locomotive Operations

When inactive, locomotives are housed in a locomotive depot (UK term) or engine house (US). In engine facilities, or a Traction Maintenance Depot, locomotives are cleaned, repaired, etc. Decommissioned locomotives are sometimes used to heat passenger cars and defrost railroad switches in winter. After this period, locomotives (and other rail vehicles) are turned into scrap or are left to rust in a train depot. Some end up in railway museums or are bought by railway preservation groups.

Background Factors (Feasibility)

Each transport system represents a contribution to a country's infrastructure, and as such must make economic sense or eventually close. From this, each has a particular role or roles. These may change with time but they affect the specifications of each particular system.

Extrinsic Factors

Rail transport systems are built into the landscape, including both the physical geography (hills, valleys, etc.) and the human geography (location of settlements). The rail transport system may in turn feedback into the human geography.

Physical Geography

The permanent way of a system must pass through the geography and geology of its region. This may be flat or mountainous, may include obstacles such as water and mountains. These determine, in part, the intrinsic nature of the system. The slope at which trains run must also be calculated correctly. In this stage, it is decided where tunnels pass.

Human Geography

Rail transport systems affect the human geography. Large cities (such as Nairobi) may be founded by a railroad passing through. Historically, when a station has been built outside the town or city it is intended to serve, that town has expanded to include the station, or buildings (especially Inns) sprung up near the station. The existence of a station may increase the number of commuters who live in a town or village and so cause it to become a dormitory town. The transcontinental railroad was a large factor in American colonization of the Western frontier. China's railroad expansion into Tibet may have similar consequences.

Historical Factors

Rail transport systems are often used for purposes they were not designed for, but have evolved into due to changes in human geography. Politics can play a large part in decisions about railways, such as the Beeching Axe. In the UK, building or rebuilding a railway usually requires an Act of Parliament.

Ownership

Traditionally, the infrastructure and rolling stock are owned and operated by the same company. This has often been by a national railway, while other companies have had private railways. Since the 1980s, there has been an increasing tendency to split up railway companies, with separate companies owning the stock from those owning the infrastructure, particularly in Europe, where this is required by the European Union. This has allowed open access by any train operator to any portion of the European railway network.

Financing

The main source of income for railway companies is from ticket revenue (for passenger transport) and shipment fees for cargo. Discounts and monthly passes are sometimes available for frequent travellers. Freight revenue may be sold per container slot or for a whole train. Sometimes, the shipper owns the cars and only rents the haulage. For passenger transport, advertisement income can be significant.

Government may choose to give subsidies to rail operation, since rail transport has fewer externalities than other dominant modes of transport. If the railway company is state-owned, the state may simply provide direct subsidies in exchange for an increased production.

If operations have been privatized, several options are available. Some countries have a system where the infrastructure is owned by a government agency or company—with open access to the tracks for any company that meets safety requirements.

In such cases, the state may choose to provide the tracks free of charge, or for a fee that does not cover all costs.

This is seen as analogous to the government providing free access to roads. For passenger operations, a direct subsidy may be paid to a public-owned operator, or public service obligation tender may be helt, and a time-limited contract awarded to the lowest bidder.

Safety

Rail transport is one of the safest forms of land travel. Trains can travel at very high speed, but they are heavy, are unable to deviate from the track and require a great distance to stop. Possible accidents include derailment (jumping the track), a head-on collision with another train and collision with an automobile or other vehicle at a level crossings. The latter accounts for the majority of rail accidents and casualties. The most important safety measures are railway signalling and gates or grade separation at crossings. Train whistles warn of the presence of a train, while trackside signals maintain the distances between trains.

Impact, Energy

Rail transport is an energy-efficient but capital-intensive means of mechanized land transport. The tracks provide smooth and hard surfaces on which the wheels of the train can roll with a minimum of friction. As an example, a typical modern wagon can hold up to 113 tonnes of freight on two four-wheel bogies. The contact area between each wheel and the rail is a strip no more than a few millimetres wide, which minimizes friction. The track distributes the weight of the train evenly, allowing significantly greater loads per axle and wheel than in road transport, leading to less wear and tear on the permanent way. This can save energy compared with other forms of transportation, such as road transport, which depends on the friction between rubber tires and the road. Trains have a small frontal area in relation to the load they are carrying, which reduces air resistance and thus energy usage, although this does not reduce the effects of side winds.

In addition, the presence of track guiding the wheels allows for very long trains to be pulled by one or a few engines, even around curves, which allows for economies of scale in energy use; by contrast, in road transport, more than two articulations causes fishtailing and makes the vehicle unsafe.

Usage

Due to these benefits, rail transport is a major form of passenger and freight transport in many countries. In India, China, South Korea and Japan, many millions use trains as regular transport. It is widespread in European countries. Freight rail transport is widespread and heavily used in North America, but intercity passenger rail transport on that continent is relatively scarce outside the Northeast Corridor.

Africa and South America have some extensive networks such as in South Africa, Northern Africa and Argentina; but some railway on these continents are isolated lines connecting two places. Australia has a generally sparse network befitting its population density, but has some areas with significant networks, especially in the southeast.

In addition to the previously existing east-west transcontinental line in Australia, a line from north to south has been constructed. The highest railway in the world is the line to Lhasa, in Tibet, partly running over permafrost territory.

The western Europe region has the highest railway density in the world, and has many individual trains which operate through several countries despite technical and organizational differences in each national network.

Of 236 countries and dependencies globally, 143 have rail transport (including several with very little), of which about 90 have passenger services.

A train consists of one or more connected vehicles that run on the rails. Propulsion is commonly provided by a locomotive, that hauls a series of unpowered cars, that can carry passengers or freight. The locomotive can be powered by steam, diesel or by electricity supplied by trackside systems. Alternatively, some or all the cars can be powered, known as a multiple unit. Also, a train can be powered by horses, cables, gravity, pneumatics and gas turbines. Railed vehicles move with much less friction than rubber tires on paved roads, making trains more energy efficient, though not as efficient as ships.

Intercity trains are long-haul services connecting cities; modern high-speed rail is capable of speeds up to 350 km/h (220 mph), but this requires specially built track. Regional and commuter trains feed cities from suburbs and surrounding areas, while intra-urban transport is performed by high-capacity tramways and rapid transits, often making up the backbone of a city's public transport. Freight trains traditionally used box cars, requiring manual loading and unloading of the cargo. Since the 1960s, container trains have become the dominant solution for general freight, while large quantities of bulk are transported by dedicated trains.

Road

Road transport (British English) or road transportation (American English) is transport on roads of passengers or goods. A hybrid of road transport and ship transport is the historic horse-drawn boat.

History

The first methods of road transport were horses, oxen or even humans carrying goods over dirt tracks that often followed game trails. As commerce increased, the tracks were often flattened or widened to accommodate the activities. Later, the travels, a frame used to drag loads, was developed.

The wheel came still later, probably preceded by the use of logs as rollers. Early stone-paved roads were built in Mesopotamia and the Indus Valley Civilization. The Persians later built a network of Royal Roads across their empire. With the advent of the Roman Empire, there was a need for armies to be able to travel quickly from one area to another, and the roads that existed were often muddy, which greatly delayed the movement of large masses of troops.

To resolve this issue, the Romans built great roads. The Roman roads used deep roadbeds of crushed stone as an underlying layer to ensure that they kept dry, as the water would flow out from the crushed stone, instead of becoming mud in clay soils. The Islamic Caliphate later built tar-paved roads in Baghdad.

During the Industrial Revolution, and because of the increased commerce that came with it, improved roadways became imperative. The problem was rain combined with dirt roads created commerce-miring mud. John Loudon McAdam (1756-1836) designed the first modern highways.

He developed an inexpensive paving material of soil and stone aggregate (known as macadam), and he embanked roads a few feet higher than the surrounding terrain to cause water to drain away from the surface. At the same time, Thomas Telford, made substantial advances in the engineering of new roads and the construction of bridges, particularly, the London to Holyhead road.

Various systems had been developed over centuries to reduce bogging and dust in cities, including cobblestones and wooden paving. Tar-bound macadam (tarmac) was applied to macadam roads towards the end of the 19th century in cities such as Paris. In the early 20th century tarmac and concrete paving were extended into the countryside.

Transportation

Transport on roads can be roughly grouped into two categories: transportation of goods and transportation of people. In many countries licencing requirements and safety regulations ensure a separation of the two industries.

The nature of road transportation of goods depends, apart from the degree of development of the local infrastructure, on the distance the goods are transported by road, the weight and volume of the individual shipment and the type of goods transported. For short distances and light, small shipments a van or pickup truck may be used. For large shipments even if less than a full truckload (Less than truckload) a truck is more appropriate. In some countries cargo is transported by road in horse-drawn carriages, donkey carts or other non-motorized mode. Delivery services are sometimes considered a separate category from cargo transport. In many places fast food is transported on roads by various types of vehicles. For inner city delivery of small packages and documents bike couriers are quite common.

People (Passengers) are transported on roads either in individual cars or automobiles or in mass transit/public transport by bus/Coach (vehicle). Special modes of individual transport by road like rikshas or velotaxis may also be locally available.

Trucking and Hauling

Trucking companies (AE) or haulers/hauliers (BE) accept cargo for road transportation. Truck drivers operate either independently working directly for the client or through freight carriers or shipping agents. Some big companies (e.g. grocery store chains) operate their own internal trucking operations.

In the U.S. many truckers own their truck (rig), and are known as owner-operators. Some road transportation is done on regular routes or for only one consignee per run, while others transport goods from many different loading stations/shippers to various consignees. On some long runs only cargo for one lag of the route (to) is known when the cargo is loaded. Truckers may have to wait at the destination for the return cargo (from).

A Bill of Lading issued by the shipper provides the basic document for road freight. On cross-border transportation the trucker will present the cargo and documentation provided by the shipper to customs for inspection.

To avoid accidents caused by fatigue, truckers have to keep to strict rules for drivetime and required rest periods. Known in the U.S. as hours of service, and in the E.U. as drivers working hours. Tachographs record the times the vehicle is in motion and stopped. Some companies use two drivers per truck to ensure uninterrupted transportation; with one driver resting or sleeping in a bunk in the back of the cab while the other is driving. Truck drivers often need special licenses to drive, known in the U.S. as a commercial driver's license. In the U.K. a Large Goods Vehicle license is required.

For transport of hazardous materials truckers need a licence, which usually requires them to pass an exam (e.g. in the EU). They have to make sure they affix proper labels for the respective hazard(s) to their vehicle. Liquid goods are transported by road in tank trucks (AE) or tanker lorries (BE) (also road-tankers) or special tankcontainers for intermodal transport. For unpackaged goods and liquids weigh stations confirm weight after loading and before delivery. For transportation of live animals special requirements have to be met in many countries to prevent cruelty to animals. For fresh and frozen goods refrigerator trucks or reefer (container)s are used.

In Australia road trains replace rail transport for goods on routes throughout the centre of the country. B-doubles and semi-trailers are used in urban areas because of their smaller size. Low-loader or flat-bed trailers are used to haul containers.

Modern Roads

Today roadways are principally asphalt or concrete. Both are based on McAdam's concept of stone aggregate in a binder, asphalt cement or Portland cement respectively. Asphalt is known as a flexible pavement, one which slowly will "flow" under the pounding of traffic. Concrete is a rigid pavement, which can take heavier loads but is more expensive and requires more carefully prepared subbase. So, generally, major roads are concrete and local roads are asphalt. Often concrete roads are covered with a thin layer of asphalt to create a wearing surface.

Modern pavements are designed for heavier vehicle loads and faster speeds, requiring thicker slabs and deeper subbase. Subbase is the layer or successive layers of stone, gravel and sand supporting the pavement. It is needed to spread out the slab load bearing on the underlying soil and to conduct away any water getting under the slabs. Water will undermine a pavement over time, so much of pavement and pavement joint design are meant to minimize the amount of water getting and staying under the slabs.

Shoulders are also an integral part of highway design. They are multipurpose; they can provide a margin of side clearance, a refuge for incapacitated vehicles, an emergency lane, and parking space. They also serve a design purpose, and that is to prevent water from percolating into the soil near the main pavement's edge. Shoulder pavement is designed to a lower standard than the pavement in the traveled way and won't hold up as well to traffic. (Which is why driving on the shoulder is generally prohibited.)

Pavement technology is still evolving, albeit in not easily noticed increments. For instance, chemical additives in the pavement mix make the pavement more weather resistant, grooving and other surface treatments improve resistance to skidding and hydroplaning, and joint seals which were once tar are now made of low maintenance neoprene.

Traffic Control

Nearly all roadways are built with devices meant to control traffic. Most notable to the motorist are those meant to communicate directly with the driver. Broadly, these fall into three categories: signs, signals or pavement markings. They help the driver navigate; they assign the right-of-way at intersections; they indicate laws such as speed limits and parking regulations; they advise of potential hazards; they indicate passing and no passing zones; and otherwise deliver information and to assure traffic is orderly and safe.

200 years ago these devices were signs, nearly all informal. In the late 19th century signals began to appear in the biggest cities at a few highly congested intersections. They were manually operated, and consisted of semaphores, flags or paddles, or in some cases coloured electric lights, all modelled on railroad signals. In the 20th century signals were automated, at first with electromechanical devices and later with computers. Signals can be quite sophisticated: with vehicle sensors embedded in the pavement, the signal can control and choreograph the turning movements of heavy traffic in the most complex of intersections. In the 1920s traffic engineers learned how to coordinate signals along a thoroughfare to increase its speeds and volumes. In the 1980s, with computers, similar coordination of whole networks became possible.

In the 1920s pavement markings were introduced. Initially they were used to indicate the road's centerline. Soon after they were coded with information to aid motorists in passing safely. Later, with multi-lane roads they were used to define lanes. Other uses, such as indicating permitted turning movements and pedestrian crossings soon followed.

In the 20th century traffic control devices were standardized. Before then every locality decided on what its devices would look like and where they would be applied. This could be confusing, especially to traffic from outside the locality. In the United States standardization was first taken at the state level, and late in the century at the federal level. Each country has a Manual of Uniform Traffic Control Devices (MUTCD) and there are efforts to blend them into a worldwide standard.

Besides signals signs and markings, other forms of traffic control are designed and built into the roadway. For instance, curbs and rumble strips can be used to keep traffic in a given lane and median barriers can prevent left turns and even U-turns.

Toll Roads

Early toll roads were usually built by private companies under a government franchise. They typically paralleled or replaced routes already with some volume of commerce, hoping the improved road would divert enough traffic to make the enterprise profitable. Plank roads were particularly attractive as they greatly reduced rolling resistance and mitigated the problem of getting mired in mud. Another improvement, better grading to lessen the steepness of the worst stretches, allowed draft animals to haul heavier loads.

A *toll road* in the United States is often called a *turnpike.* The term *turnpike* probably originated from the gate, often a simple pike, which blocked passage until the fare was paid at a *toll house* (or *toll booth* in current terminology).

When the toll was paid the pike, which was mounted on a swivel, was turned to allow the vehicle to pass. Tolls were usually based on the type of cargo being transported, not the type of vehicle. The practice of selecting routes so as to avoid tolls is called shunpiking. This may be simply to avoid the expense, as a form of economic protest (or boycott), or simply to seek a road less traveled as a bucolic interlude. Companies were formed to build, improve, and maintain a particular section of roadway, and tolls were collected from users to finance the enterprise. The enterprise was usually named to indicate the locale of its roadway, often including the name of one of both of the termini. The word *turnpike* came into common use in the names of these roadways and companies, and is essentially used interchangeably with *toll road* in current terminology.

In the United States, toll roads began with the Lancaster Turnpike in the 1790s, within Pennsylvania, connecting Philadelphia and Lancaster.

In New York State, the Great Western Turnpike was started in Albany in 1799 and eventually extended, by several alternate routes, to near what is now Syracuse, New York.

Toll roads peaked in the mid 19th century, and by the turn of the twentieth century most toll roads were taken over by state highway departments. The demise of this early toll road era was due to the rise of canals and railroads, which were more efficient (and thus cheaper) in moving freight over long distances. Roads wouldn't again be competitive with rails and barges until the first half of the 20th century when the internal combustion engine replaces draft animals as the source of motive power.

With the development, mass production, and popular embrace of the automobile, faster and higher capacity roads were needed. In the 1920s limited access highways appeared. Their main characteristics were dual roadways with access points limited to (but not always) grade-separated interchanges. Their dual roadways allowed high volumes of traffic, the need for no or few traffic lights along with relatively gentle grades and curves allowed higher speeds.

The first limited access highways were *Parkways*, so called because of their often park-like landscaping and, in the metropolitan New York City area, they connected the region's system of parks. When the German Autobahns built in the 1930s introduced higher design standards and speeds, road planners and road-builders in the United States started developing and building toll roads to similar high standards. The Pennsylvania Turnpike, which largely followed the path of a partially-built railroad, was the first, opening in 1940. After 1940 with the Pennsylvania Turnpike, toll roads saw a resurgence, this time to fund limited access highways. In the late 1940s and early 1950s, after World War II interrupted the evolution of the highway, the US resumed building toll roads. They were to still higher standards and one road, the New York State Thruway, had standards that became the prototype for the U.S. Interstate Highway System. Several other major toll-roads which connected with the Pennsylvania Turnpike were established before the creation of the Interstate Highway System. These were the Indiana Toll Road, Ohio Turnpike, and New Jersey Turnpike.

US Interstate Highway System

In the United States, beginning in 1956, Dwight D. Eisenhower National System of Interstate and Defence Highways, commonly called the Interstate Highway System was built. It uses 12 foot (3.65m) lanes, wide medians, a maximum of 4% grade, and full access control, though many sections don't meet these standards due to older construction or constraints. This system created a continental-sized network meant to connect every population centre of 50,000 people or more.

By 1956, most limited access highways in the eastern United States were toll roads. In that year, the federal Interstate highway program was established, funding non-toll roads with 90% federal dollars and 10% state match, giving little incentive for states to expand their turnpike system. Funding rules initially restricted collections of tolls on newly funded roadways, bridges, and tunnels. In some situations, expansion or rebuilding of a toll facility using Interstate Highway Program funding resulted in the removal of existing tolls. This occurred in Virginia on Interstate 64 at the Hampton Roads Bridge-Tunnel when a second parallel roadway to the regional 1958 bridge-tunnel was completed in 1976.

Since the completion of the initial portion of the interstate highway system, regulations were changed, and portions of toll facilities have been added to the system. Some states are again looking at toll financing for new roads and maintenance, to supplement limited federal funding. In some areas, new road projects have been completed with public-private partnerships funded by tolls, such as the Pocahontas Parkway (I-895) near Richmond, Virginia.

Pneumatic Tires

As the horse-drawn carriage was replaced by the car and lorry or truck, and speeds increased, the need for smoother roads and less vertical displacement became more apparent, and pneumatic tires were developed to decrease the apparent roughness. Wagon and carriage wheels, made of wood, had a tire in the form of an iron strip that kept the wheel from wearing out quickly. Pneumatic tires, which had a larger footprint than iron tires, also were less likely to get bogged down in the mud on unpaved roads.

Road Transport and the Environment

The environmental impact of roads includes the local effects of highways (public roads) such as on noise, water pollution, habitat destruction/disturbance and local air quality; and the wider effects including climate change from vehicle emissions. The design, construction and management of roads, parking and other related facilities as well as the design and regulation of vehicles can change the impacts to varying degrees.

Air Quality

Roads can have both negative and positive effects on air quality.

Negative Impacts

Air pollution from Motor vehicle emissions can occur wherever vehicles are used and are of particular concern in congested city street conditions and other low speed circumstances. Emissions include particulate emissions from diesel engines, NO_x, volatile organic compounds, Carbon monoxide and various other hazardous air pollutants including benzene. Concentrations of air pollutants and adverse respiratory health effects are greater near the road than at some distance away from the road. Road dust kicked up by vehicles may trigger allergic reactions. Carbon dioxide is non-toxic to humans but is a major greenhouse gas and motor vehicle emissions are an important contributor to the growth of CO_2 concentrations in the atmosphere and therefore to global warming.

Positive Impacts

The construction of new roads which divert traffic from built-up areas can deliver improved air quality to the areas relieved of a significant amount of traffic. The *Environmental and Social Impact Assessment Study* carried out for the development of the Tirana Outer Ring Road estimated that it would result in improved air quality in Tirana city centre.

Noise

Motor vehicle traffic on roads will generate noise.

Negative Impacts

Road noise can be a nuisance if it impinges on population centres, especially for roads at higher operating speeds, near intersections and on uphill sections. Noise health effects can be expected in such locations from road systems used by large numbers of motor vehicles. Noise mitigation strategies exist to reduce sound levels at nearby sensitive receptors. The idea that road design could be influenced by acoustical engineering considerations first arose about 1973.

Speed bumps, which are usually deployed in built-up areas, can increase noise pollution, especially if large vehicles use the road and particularly at night.

Positive Impacts

New roads can divert traffic away from population centres thus relieving the noise pollution. A new road scheme planned in Shropshire, UK promises to reduce traffic noise in Shrewsbury town centre.

Water Pollution

Urban runoff from roads and other impervious surfaces is a major source of water pollution. Rainwater and snowmelt running off of roads tends to pick up gasoline, motor oil, heavy metals, trash and other pollutants. Road runoff is a major source of nickel, copper, zinc, cadmium, lead and polycyclic aromatic hydrocarbons (PAHs), which are created as combustion byproducts of gasoline and other fossil fuels. De-icing chemicals and sand can run off into roadsides, contaminate groundwater and pollute surface waters. Road salts (primarily chlorides of sodium, calcium or magnesium) can be toxic to sensitive plants and animals. Sand can alter stream bed environments, causing stress for the plants and animals that live there.

Habitat Fragmentation

Roads can act as barriers or filters to animal movement and lead to habitat fragmentation. Many species will not cross the open space created by a road due to the threat of predation and roads also cause increased animal mortality from traffic. This barrier effect can prevent species from migrating and recolonising areas where the species has gone locally extinct as well as restricting access to seasonally available or widely scattered resources.

Habitat fragmentation may also divide large continuous populations into smaller more isolated populations. These smaller populations are more vulnerable to genetic drift, inbreeding depression and an increased risk of population decline and extinction.

By subsector, road transport is the largest contributor to global warming (74% of total emissions from transport).

Water

Water transport is the process of transport a watercraft, such as a barge, boat, ship or sailboat, makes over a body of water, such as a sea, ocean, lake, canal or river. The need for buoyancy unites watercraft, and makes the hull a dominant aspect of its construction, maintenance and appearance.

In the 1800s the first steam ships were developed, using a steam engine to drive a paddle wheel or propeller to move the ship. The steam was produced using wood or coal. Now most ships have an engine using a slightly refined type of petroleum called bunker fuel. Some ships, such as submarines, use nuclear power to produce the steam. Recreational or educational craft still use wind power, while some smaller craft use internal combustion engines to drive one or more propellers, or in the case of jet boats, an inboard water jet. In shallow draft areas, hovercraft are propelled by large pusher-prop fans.

Although slow, modern sea transport is a highly effective method of transporting large quantities of non-perishable goods. Commercial vessels, nearly 35,000 in number, carried 7.4 billion tons of cargo in 2007. Transport by water is significantly less costly than air transport for trans-continental shipping; short sea shipping and ferries remain viable in coastal areas.

Ship Transport

Ship transport is watercraft carrying people (passengers) or goods (cargo). Sea transport has been the largest carrier of freight throughout recorded history. Although the importance of sea travel for passengers has decreased due to aviation, it is effective for short trips and pleasure cruises. Transport by water is cheaper than transport by air. Ship transport can be over any distance by boat, ship, sailboat or barge, over oceans and lakes, through canals or along rivers. Shipping may be for commerce, recreation or the military. Virtually any material that can be moved, can be moved by water, however water transport becomes impractical when material delivery is highly time-critical. "General cargo" is goods packaged in boxes, cases, pallets, and barrels. Containerization revolutionized ship transport in the 1960s. When a cargo is carried in more than one mode, it is intermodal or co-modal.

Merchant Shipping

A nation's shipping fleet (merchant navy, merchant marine, merchant fleet) consists of the ships operated by civilian crews to transport passengers or cargo. Professionals are merchant seaman, merchant sailor, and merchant mariner, or simply seaman, sailor, or mariner. The terms "seaman" or "sailor" may refer to a member of a country's navy.

According to the 2005 CIA World Factbook, the world total number of merchant ships of 1,000 Gross Register Tons or over was 30,936.

Professional mariners. A ship's complement can be divided into four categories: the deck department, the engineering department, the steward's department, and other.

Deck Department

For more details on this topic, see Deck department.

Officer positions in the deck department include but not limited to: Master and his Chief, Second, and Third officers. The official classifications for unlicensed members of the deck department are Able Seaman and Ordinary Seaman.

A common deck crew for a ship includes:

- (1) Chief Officer/Chief Mate
- (1) Second Officer/Second Mate
- (1) Third Officer/Third Mate
- (1) Boatswain
- (2-6) Able Seamen
- (0-2) Ordinary Seamen.

A deck cadet is person who is carrying out mandatory seatime to achieve their officer of the watch certificate. Their time onboard is spent learning the operations and tasks of everyday life on a merchant vessel.

Engineering Department

A ship's engineering department consists of the members of a ship's crew that operate and maintain the propulsion and other systems on board the vessel. Marine Engineering staff also deal with the "Hotel" facilities on board, notably the sewage, lighting, air conditioning and water systems. They deal with bulk fuel transfers, and require training in firefighting and first aid, as well as in dealing with the ship's boats and other nautical tasks-especially with cargo loading/ discharging gear and safety systems, though the specific cargo discharge function remains the responsibility of deck officers and deck workers. On LPG and LNG tankers however, a cargo engineer works with the deck department during cargo operations, as well as being a watchkeeping engineer.

A common Engineering crew for a ship includes:

- (1) Chief Engineer
- (1) Second Engineer/First Assistant Engineer
- (1) Third Engineer/Second Assistant Engineer

- (1-2) Fourth Engineer/Third Assistant Engineer
- (0-2) Fifth Engineer/Junior Engineer
- (1-3) Oiler (unlicensed qualified rating)
- (0-3) Greaser/s (unlicensed qualified rating)
- (1-5) Entry-level rating (such as Wiper (occupation), Utilityman, etc.).

Many American ships also carry a Qualified Member of the Engine Department. Other possible positions include Motorman, Machinist, Electrician, Refrigeration Engineer, and Tankerman. Engine Cadets are trainee engineers who are completing sea time necessary before they can obtain a watchkeeping license.

Steward's Department

A typical Steward's department for a cargo ship would be composed of a Chief Steward, a Chief Cook, and a Steward's Assistant. All three positions are typically filled by unlicensed personnel. The chief steward directs, instructs, and assigns personnel performing such functions as preparing and serving meals; cleaning and maintaining officers' quarters and steward department areas; and receiving, issuing, and inventorying stores.

On large passenger vessels, the Catering Department is headed by the Chief Purser and managed by assistant pursers. Although they enjoy the benefits of having officer rank, they generally progress through the ranks to become pursers. Under the pursers are the department heads-such as chief cook, head waiter, head barman etc. They are responsible for the administration of their own areas.

The chief steward also plans menus; compiles supply, overtime, and cost control records. May requisition or purchase stores and equipment. May bake bread, rolls, cakes, pies, and pastries. A chief steward's duties may overlap with those of the Steward's Assistant, the Chief Cook, and other Steward's Department crewmembers. In the United States Merchant Marine, in order to be occupied as a chief steward a person has to have a Merchant Mariner's Document issued by the United States Coast Guard. Because of international conventions and agreements, all chief cooks who sail internationally are similarly documented by their respective countries.

Other Departments

Staff officer positions on a ship, including Junior Assistant Purser, Senior Assistant Purser, Purser, Chief Purser, Medical Doctor,

Professional Nurse, Marine Physician Assistant, and Hospital Corpsman, are considered administrative positions and are therefore regulated by Certificates of Registry issued by the United States Coast Guard. Pilots are also merchant marine officers and are licensed by the Coast Guard.

Formerly, there was also a radio department, headed by a chief radio officer and supported by a number of radio officers. Since the introduction of GMDSS (Satellite communications) and the subsequent exemptions from carrying radio officers if the vessel is so equipped, this department has fallen away, although many ships do still carry specialist radio officers, particularly passenger vessels. Many radio officers became 'electro-technical officers', and transferred into the engineering department.

Life at Sea

Mariners live much of their life spent beyond the reach of land. They face sometimes dangerous conditions at sea. Yet men and women still go to sea. For some, the attraction is a life unencumbered with the restraints of life ashore. Sea-going adventure and a chance to see the world also appeal to many seafarers. Whatever the calling, those who live and work at sea invariably confront social isolation.

Findings by the Seafarer's International Research Centre indicate a leading cause of mariners leaving the industry is "almost invariably because they want to be with their families." U.S. merchant ships typically do not allow family members to accompany seafarers on voyages. Industry experts increasingly recognize isolation, stress, and fatigue as occupational hazards. Advocacy groups such as International Labour Organization, a United Nations agency, and the Nautical Institute are seeking improved international standards for mariners.

Ocean voyages are steeped in routine. Maritime tradition dictates that each day be divided into six four-hour periods. Three groups of watchkeepers from the engine and deck departments work four hours on then have eight hours off watchkeeping. However there are many overtime jobs to be done daily. This cycle repeats endlessly, 24 hours a day while the ship is at sea. Members of the steward department typically are day workers who put in at least eight-hour shifts. Operations at sea, including repairs, safeguarding against piracy, securing cargo, underway replenishment, and other duties provide opportunities for overtime work. Service aboard ships typically extends for months at a time, followed by protracted shore leave. However, some seamen secure jobs on ships they like and stay aboard for years.

In rare cases, veteran mariners choose never to go ashore when in port. Further, the often quick turnaround of many modern ships, spending only a matter of hours in port, limits a seafarer's free-time ashore. Moreover, some foreign seamen entering U.S. Ports from a watchlist of 25 high-risk countries face restrictions on shore leave due to security concerns in a post 9/11 environment. However, shore leave restrictions while in U.S. Ports impact American seamen as well. For example, the International Organization of Masters, Mates & Pilots notes a trend of U.S. Shipping terminal operators restricting seamen from travelling from the ship to the terminal gate. Further, in cases where transit is allowed, special "security fees" are at times assessed.

Such restrictions on shore leave coupled with reduced time in port by many ships translate into longer periods at sea. Mariners report that extended periods at sea living and working with shipmates who for the most part are strangers takes getting used to. At the same time, there is an opportunity to meet people from other ethnic and cultural backgrounds. Recreational opportunities have improved aboard some U.S. Ships, which may feature gyms and day rooms for watching movies, swapping sea stories, and other activities. And in some cases, especially tankers, it is made possible for a mariner to be accompanied by members of his family. However, a mariner's off duty time is largely a solitary affair, pursuing hobbies, reading, writing letters, and sleeping.

On modern ocean going vessels, typically registered with a flag of convenience, life has changed immensely in the last 20 years. Most large vessels include a gym and often a swimming pool for use by the crew. Since the Exxon Valdez incident, the focus of leisure time activity has shifted from having officer and crew bars, to simply having lounge-style areas where officers or crew can sit to watch movies. With many companies now providing TVs and DVD players in cabins, and enforcing strict smoking policies, it is not surprising that the bar is now a much quieter place on most ships. In some instances games consoles are provided for the officers and crew. The officers enjoy a much higher standard of living on board ocean going vessels. Crews are generally poorly paid, poorly qualified and have to complete contracts of approx 9 months before returning home on leave. They often come from countries where the average industrial wage is still very low, such as the Philippines or India. Officers however, come from all over the world and it is not uncommon to mix the nationality of the officers on board ships. Officers are often the recipients of university degrees and have completed vast amounts of training in order to reach their rank. Officers benefit on board by having larger, more comfortable cabins, table service for their meals, etc. Contracts average at the 4 month mark for officers, with generous leave. Most Ocean going vessels now operate an Unmanned Engineroom System allowing engineers to work days only. The engine room is computer controlled by night, although the duty engineer will make inspections during unmanned operation. Engineers work in a hot, humid, noisy atmosphere. Communication in the engineroom is therefore by hand signals and lip-reading, and good teamwork often stands in place of any communication at all.

Ships and Watercraft

Ships and other watercraft are used for ship transport. Types can be distinguished by propulsion, size or cargo type. Recreational or educational craft still use wind power, while some smaller craft use internal combustion engines to drive one or more propellers, or in the case of jet boats, an inboard water jet. In shallow draft areas, such as the Everglades, some craft, such as the hovercraft, are propelled by large pusher-prop fans.

Most modern merchant ships can be placed in one of a few categories, such as:

- Bulk carriers, such as the Sabrina I seen here, are cargo ships used to transport bulk cargo items such as ore or food staples (rice, grain, etc.) and similar cargo. It can be recognized by the large box-like hatches on its deck, designed to slide outboard for loading. A bulk carrier could be either dry or wet. Most lakes are too small to accommodate bulk ships, but a large fleet of lake freighters has been plying the Great Lakes and St. Lawrence Seaway of North America for over a century.
- Container ships are cargo ships that carry their entire load in truck-size containers, in a technique called containerization. They form a common means of commercial intermodal freight transport. Informally known as "box boats," they carry the majority of the world's dry cargo. Most container ships are propelled by diesel engines, and have crews of between 10 and 30 people. They generally have a large accommodation block at the stern, directly above the engine room.
- Tankers are cargo ships for the transport of fluids, such as crude oil, petroleum products, liquefied petroleum gas, liquefied natural gas and chemicals, also vegetable oils, wine and other food-the tanker sector comprises one third of the world tonnage.

- Reefer ships are cargo ships typically used to transport perishable commodities which require temperature-controlled transportation, mostly fruits, meat, fish, vegetables, dairy products and other foodstuffs.
- Roll-on/roll-off ships, such as the Chi-Cheemaun, are cargo ships designed to carry wheeled cargo such as automobiles, trailers or railway carriages. RORO (ro/ro) vessels have built-in ramps which allow the cargo to be efficiently "rolled on" and "rolled off" the vessel when in port. While smaller ferries that operate across rivers and other short distances still often have built-in ramps, the term RORO is generally reserved for larger ocean-going vessels.
- Coastal trading vessels, also known as coasters, are shallow-hulled ships used for trade between locations on the same island or continent. Their shallow hulls mean that they can get through reefs where sea-going ships usually cannot (sea-going ships have a very deep hull for supplies and trade etc.).
- Ferries are a form of transport, usually a boat or ship, but also other forms, carrying (or *ferrying*) passengers and sometimes their vehicles. Ferries are also used to transport freight (in lorries and sometimes unpowered freight containers) and even railroad cars. Most ferries operate on regular, frequent, return services. A foot-passenger ferry with many stops, such as in Venice, is sometimes called a waterbus or water taxi. Ferries form a part of the public transport systems of many waterside cities and islands, allowing direct transit between points at a capital cost much lower than bridges or tunnels. Many of the ferries operating in Northern European waters are ro/ro ships.
- Cruise ships are passenger ships used for pleasure voyages, where the voyage itself and the ship's amenities are considered an essential part of the experience. Cruising has become a major part of the tourism industry, with millions of passengers each year as of 2006. The industry's rapid growth has seen nine or more newly built ships catering to a North American clientele added every year since 2001, as well as others servicing European clientele. Smaller markets such as the Asia-Pacific region are generally serviced by older tonnage displaced by new ships introduced into the high growth areas. On the Baltic sea this market is served by cruiseferries.

- Cable layer is a deep-sea vessel designed and used to lay underwater cables for telecommunications, electricity, and such. A large superstructure, and one or more spools that feed off the transom distinguish it.
- A tugboat is a boat used to manoeuvre, primarily by towing or pushing other vessels in harbours, over the open sea or through rivers and canals. They are also used to tow barges, disabled ships, or other equipment like towboats.
- A dredger (sometimes also called a dredge) is a ship used to excavate in shallow seas or fresh water areas with the purpose of gathering up bottom sediments and disposing of them at a different location.
- A barge is a flat-bottomed boat, built mainly for river and canal transport of heavy goods. Most barges are not self-propelled and need to be moved by tugboats towing or towboats pushing them. Barges on canals (towed by draft animals on an adjacent towpath) contended with the railway in the early industrial revolution but were outcompeted in the carriage of high value items due to the higher speed, falling costs, and route flexibility of rail transport.

Ships that fall outside these categories include Semi-submersible heavy-lift ships or OHGC.

Typical In-transit Times

A cargo ship sailing from a European port to a US one will typically take 10–12 days depending on water currents and other factors.

Ship Transport Infrastructure

For a port to efficiently send and receive cargo, it requires infrastructure. Harbours, seaports and marinas host watercraft, and consist of components such as piers, wharfs, docks and roadsteads.

Port

A port is a location on a coast or shore containing one or more harbours where ships can dock and transfer people or cargo to or from land. Port locations are selected to optimize access to land and navigable water, for commercial demand, and for shelter from wind and waves. Ports with deeper water are rarer, but can handle larger, more economical ships. Since ports throughout history handled every kind of traffic, support and storage facilities vary widely, may extend for miles, and dominate the local economy. Some ports have an important, perhaps exclusively military role.

Distribution

Ports often have cargo-handling equipment, such as cranes (operated by longshoremen) and forklifts for use in loading ships, which may be provided by private interests or public bodies. Often, canneries or other processing facilities will be located nearby. Some ports feature canals, which allow ships further movement inland. Access to intermodal transportation, such as trains and trucks, are critical to a port, so that passengers and cargo can also move further inland beyond the port area. Ports with international traffic have customs facilities. Harbour pilots and tugboats may manoeuvre large ships in tight quarters when near docks.

Port Types

The terms "port" and "seaport" are used for different types of port facilities that handle ocean-going vessels, and river port is used for river traffic, such as barges and other shallow-draft vessels. Some ports on a lake, river, or canal have access to a sea or ocean, and are sometimes called "inland ports".

A fishing port is a port or harbour facility for landing and distributing fish. It may be a recreational facility, but it is usually commercial. A fishing port is the only port that depends on an ocean product, and depletion of fish may cause a fishing port to be uneconomical. In recent decades, regulations to save fishing stock may limit the use of a fishing port, perhaps effectively closing it.

A "dry port" is a term sometimes used to describe a yard used to place containers or conventional bulk cargo, usually connected to a seaport by rail or road.

A warm water port is where the water does not freeze in winter time. Because they are available year-round, warm water ports can be of great geopolitical or economic interest.

A seaport is further categorized as a "cruise port" or a "cargo port". Additionally, "cruise ports" are also known as a "home port" or a "port of call". The "cargo port" is also further categorized into a "bulk" or "break bulk port" or as a "container port".

A cruise home port is the port where cruise-ship passengers board (or embark) to start their cruise and also debark (or disembark) the cruise ship at the end of their cruise. It is also where the cruise ship's supplies are loaded for the cruise, which includes everything from fresh water and fuel to fruits, vegetable, champagne, and any other supplies needed for the cruise. "Cruise home ports" are a very busy place during the day the cruise ship is in port, because off-going passengers debark their baggage and on-coming passengers board the ship in addition to all the supplies being loaded. Currently, the *Cruise Capital of the World* is the Port of Miami, Florida, closely followed behind by Port Everglades, Florida and the Port of San Juan, Puerto Rico.

A port of call is an intermediate stop for a ship on its sailing itinerary, which may include up to half a dozen ports. At these ports, a cargo ship may take on supplies or fuel, as well as unloading and loading cargo. But for a cruise ship, it is their premier stop where the cruise lines take on passengers to enjoy their vacation.

Cargo ports, on the other hand, are quite different from cruise ports, because each handles very different cargo, which has to be loaded and unloaded by very different mechanical means. The port may handle one particular type of cargo or it may handle numerous cargoes, such as grains, liquid fuels, liquid chemicals, wood, automobiles, etc. Such ports are known as the "bulk" or "break bulk ports". Those ports that handle containerized cargo are known as container ports. Most cargo ports handle all sorts of cargo, but some ports are very specific as to what cargo they handle. Additionally, the individual cargo ports are divided into different operating terminals which handle the different cargoes, and are operated by different companies, also known as terminal operators or stevedores.

Access

Ports sometimes fall out of use. Rye, East Sussex was an important English port in the Middle Ages, but the coastline changed and it is now 2 miles (3.2 km) from the sea, while the ports of Ravenspurn and Dunwich have been lost to coastal erosion. Also in the United Kingdom, London, the River Thames was once an important international port, but changes in shipping methods, such as the use of containers and larger ships, put it at a disadvantage.

Other

Pipeline transport sends goods through a pipe, most commonly liquid and gases are sent, but pneumatic tubes can also send solid capsules using compressed air. For liquids/gases, any chemically stable liquid or gas can be sent through a pipeline. Short-distance systems exist for sewage, slurry, water and beer, while long-distance networks are used for petroleum and natural gas.

Cable transport is a broad mode where vehicles are pulled by cables instead of an internal power source. It is most commonly used at steep gradient. Typical solutions include aerial tramway, elevators, escalator and ski lifts; some of these are also categorized as conveyor transport.

Spaceflight is transport out of Earth's atmosphere into outer space by means of a spacecraft. While large amounts of research have gone into technology, it is rarely used except to put satellites into orbit, and conduct scientific experiments. However, man has landed on the moon, and probes have been sent to all the planets of the Solar System.

Suborbital spaceflight is the fastest of the existing and planned transport systems from a place on Earth to a distant other place on Earth. Faster transport could be achieved through part of a Low Earth orbit, or following that trajectory even faster using the propulsion of the rocket to steer it.

Elements

Infrastructure

Infrastructure is the fixed installations that allow a vehicle to operate. It consists of both a way, terminal and facilities for parking and maintenance. For rail, pipeline, road and cable transport, the entire way the vehicle travels must be built up. Air and water craft are able to avoid this, since the airway and seaway do not need to be built up. However, they require fixed infrastructure at terminals.

Terminals such as airports, ports and stations, are locations where passengers and freight can be transferred from one vehicle or mode to another. For passenger transport, terminals are integrating different modes to allow riders to interchange to take advantage of each mode's advantages. For instance, airport rail links connect airports to the city centres and suburbs. The terminals for automobiles are parking lots, while buses and coaches can operates from simple stops. For freight, terminals act as transshipment points, though some cargo is transported directly from the point of production to the point of use. The financing of infrastructure can either be public or private. Transport is often a natural monopoly and a necessity for the public; roads, and in some countries railways and airports are funded through taxation. New infrastructure projects can involve large spendings, and are often financed through debt. Many infrastructure owners therefore impose usage fees, such as landing fees at airports, or toll plazas on roads. Independent of this, authorities may impose taxes on the purchase or use of vehicles.

Vehicles

A vehicle is any non-living device that is used to move people and goods. Unlike the infrastructure, the vehicle moves along with the cargo and riders. Vehicles that do not operate on land, are usually called crafts. Unless being pulled by a cable or muscle-power, the vehicle must provide its own propulsion; this is most commonly done through a steam engine, combustion engine, electric motor, a jet engine or a rocket, though other means of propulsion also exist. Vehicles also need a system of converting the energy into movement; this is most commonly done through wheels, propellers and pressure.

Vehicles are most commonly staffed by a driver. However, some systems, such as people movers and some rapid transits, are fully automated. For passenger transport, the vehicle must have a compartment for the passengers. Simple vehicles, such as automobiles, bicycles or simple aircraft, may have one of the passengers as a driver.

Operation

Private transport is only subject to the owner of the vehicle, who operates the vehicle themselves. For public transport and freight transport, operations are done through private enterprise or by governments. The infrastructure and vehicles may be owned and operated by the same company, or they may be operated by different entities. Traditionally, many countries have had a national airline and national railway. Since the 1980s, many of these have been privatized. International shipping remains a highly competitive industry with little regulation, but ports can be public owned.

Function

Relocation of travellers and cargo are the most common uses of transport. However, other uses exist, such as the strategic and tactical relocation of armed forces during warfare, or the civilian mobility construction or emergency equipment.

Passenger

Passenger transport, or travel, is divided into public and private transport. Public is scheduled services on fixed routes, while private is vehicles that provide ad hoc services at the riders desire. The latter offers better flexibility, but has lower capacity, and a higher environmental impact. Travel may be as part of daily commuting, for business, leisure or migration.

Short-haul transport is dominated by the automobile and mass transit. The latter consists of buses in rural and small cities, supplemented with commuter rail, trams and rapid transit in larger cities. Long-haul transport involves the use of the automobile, trains, coaches and aircraft, the last of which have become predominantly used for the longest, including intercontinental, travel. Intermodal passenger transport is where a journey is performed through the use of several modes of transport; since all human transport normally starts and ends with walking, all passenger transport can be considered intermodal. Public transport may also involve the intermediate change of vehicle, within or across modes, at a transport hub, such as a bus or railway station.

Taxis and Buses can be found on both ends of Public Transport spectrum, whereas Buses remain the cheaper mode of transport but are not necessarily flexible, and Taxis being very flexible but more expensive. In the middle is Demand responsive transport offering flexibility whilst remaining affordable.

International travel may be restricted for some individuals due to legislation and visa requirements.

Freight

Freight transport, or shipping, is a key in the value chain in manufacturing. With increased specialization and globalization, production is being located further away from consumption, rapidly increasing the demand for transport. While all modes of transport are used for cargo transport, there is high differentiation between the nature of the cargo transport, in which mode is chosen. Logistics refers to the entire process of transferring products from producer to consumer, including storage, transport, transshipment, warehousing, material-handling and packaging, with associated exchange of information. Incoterm deals with the handling of payment and responsibility of risk during transport.

Containerization, with the standardization of ISO containers on all vehicles and at all ports, has revolutionized international and domestic trade, offering huge reduction in transshipment costs. Traditionally, all cargo had to be manually loaded and unloaded into the haul of any ship or car; containerization allows for automated handling and transfer between modes, and the standardized sizes allow for gains in economy of scale in vehicle operation. This has been one of the key driving factors in international trade and globalization since the 1950s.

Bulk transport is common with cargo that can be handled roughly without deterioration; typical examples are ore, coal, cereals and petroleum. Because of the uniformity of the product, mechanical handling can allow enormous quantities to be handled quickly and efficiently. The low value of the cargo combined with high volume also means that economies of scale become essential in transport, and gigantic ships and whole trains are commonly used to transport bulk. Liquid products with sufficient volume may also be transported by pipeline.

Air freight has become more common for products of high value; while less than one percent of world transport by volume is by airline, it amounts to forty percent of the value. Time has become especially important in regards to principles such as postponement and just-in-time within the value chain, resulting in a high willingness to pay for quick delivery of key components or items of high value-to-weight ratio. In addition to mail, common items send by air include electronics and fashion clothing.

History

Humans' first means of transport were walking and swimming. The domestication of animals introduces a new way to lay the burden of transport on more powerful creatures, allowing heavier loads to be hauled, or humans to ride the animals for higher speed and duration. Inventions such as the wheel and sled helped make animal transport more efficient through the introduction of vehicles. Also water transport, including rowed and sailed vessels, dates back to time immemorial, and was the only efficient way to transport large quantities or over large distances prior to the Industrial Revolution.

The first forms of road transport were horses, oxen or even humans carrying goods over dirt tracks that often followed game trails. Paved roads were built by many early civilizations, including Mesopotamia and the Indus Valley Civilization. The Persian and Roman empires built stone-paved roads to allow armies to travel quickly. Deep roadbeds of crushed stone underneath ensured that the roads kept dry. The medieval Caliphate later built tar-paved roads. The first watercraft were canoes cut out from tree trunks. Early water transport was accomplished with ships that were either rowed or used the wind for propulsion, or a combination of the two. The importance of water has led to most cities, that grew up as sites for trading, being located on rivers or at sea, ofter at the intersection of two bodies of water. Until the Industrial Revolution, transport remained slow and costly, and production and consumption were located as close to each other as feasible.

The Industrial Revolution in the 19th century saw a number of inventions fundamentally change transport. With telegraphy, communication became instant and independent of transport. The invention of the steam engine, closely followed by its application in rail transport, made land transport independent of human or animal muscles. Both speed and capacity increased rapidly, allowing specialization through manufacturing being located independent of natural resources. The 19th century also saw the development of the steam ship, that sped up global transport.

The development of the combustion engine and the automobile at the turn into the 20th century, road transport became more viable, allowing the introduction of mechanical private transport. The first highways were constructed during the 19th century with macadam. Later, tarmac and concrete became the dominant paving material. In 1903, the first controllable airplane was invented, and after World War I, it became a fast way to transport people and express goods over long distances.

After World War II, the automobile and airlines took higher shares of transport, reducing rail and water to freight and short-haul passenger. Spaceflight was launched in the 1950s, with rapid growth until the 1970s, when interest dwindled. In the 1950s, the introduction of containerization gave massive efficiency gains in freight transport, permitting globalization. International air travel became much more accessible in the 1960s, with the commercialization of the jet engine. Along with the growth in automobiles and motorways, this introduced a decline for rail and water transport. After the introduction of the Shinkansen in 1964, high-speed rail in Asia and Europe started taking passengers on long-haul routes from airlines.

Sustainable Transport

Sustainable transport (or green transport) refers to any means of transport with low impact on the environment, and includes walking and cycling, transit oriented development, green vehicles, CarSharing, and building or protecting urban transport systems that are fuel-efficient, space-saving and promote healthy lifestyles.

Sustainable transport systems make a positive contribution to the environmental, social and economic sustainability of the communities they serve. Transport systems exist to provide social and economic connections, and people quickly take up the opportunities offered by increased mobility. The advantages of increased mobility need to be weighed against the environmental, social and economic costs that transport systems pose. Transport systems have significant impacts on the environment, accounting for between 20% and 25% of world energy consumption and carbon dioxide emissions. Greenhouse gas emissions from transport are increasing at a faster rate than any other energy using sector. Road transport is also a major contributor to local air pollution and smog.

The social costs of transport include road crashes, air pollution, physical inactivity, time taken away from the family while commuting and vulnerability to fuel price increases. Many of these negative impacts fall disproportionately on those social groups who are also least likely to own and drive cars. Traffic congestion imposes economic costs by wasting people's time and by slowing the delivery of goods and services. Traditional transport planning aims to improve mobility, especially for vehicles, and may fail to adequately consider wider impacts. But the real purpose of transport is access-to work, education, goods and services, friends and family-and there are proven techniques to improve access while simultaneously reducing environmental and social impacts, and managing traffic congestion. Communities which are successfully improving the sustainability of their transport networks are doing so as part of a wider programme of creating more vibrant, livable, sustainable cities.

Definition

The term sustainable transport came into use as a logical follow-on from sustainable development, and is used to describe modes of transport, and systems of transport planning, which are consistent with wider concerns of sustainability. There are many definitions of the sustainable transport, and of the related terms sustainable transportation and sustainable mobility. One such definition, from the European Union Council of Ministers of Transport, defines a sustainable transportation system as one that:

- Allows the basic access and development needs of individuals, companies and society to be met safely and in a manner consistent with human and ecosystem health, and promotes equity within and between successive generations.

- Is Affordable, operates fairly and efficiently, offers a choice of transport mode, and supports a competitive economy, as well as balanced regional development.
- Limits emissions and waste within the planet's ability to absorb them, uses renewable resources at or below their rates of generation, and uses non-renewable resources at or below the rates of development of renewable substitutes, while minimizing the impact on the use of land and the generation of noise.

History

Most of the tools and concepts of sustainable transport were developed before the phrase was coined. Walking, the first mode of transport, is also the most sustainable. Public transport dates back at least as far as the invention of the public bus by Blaise Pascal in 1662. The first passenger tram began operation in 1807 and the first passenger rail service in 1825. Pedal bicycles date from the 1860s. These were the only personal transport choices available to most people in Western countries prior to World War II, and remain the only options for most people in the developing world. Freight was moved by human power, animal power or rail.

The post-war years brought increased wealth and a demand for much greater mobility for people and goods. The number of road vehicles in Britain increased fivefold between 1950 and 1979, with similar trends in other Western nations. Most affluent countries and cities invested heavily in bigger and better-designed roads and motorways, which were considered essential to underpin growth and prosperity. Transport planning became a branch of civil engineering and sought to design sufficient road capacity to provide for the projected level of traffic growth at acceptable levels of traffic congestion-a technique called "predict and provide". Public investment in transit, walking and cycling declined dramatically in the United States, Great Britain and Australasia, although this did not occur to the same extent in Canada or mainland Europe.

Concerns about the sustainability of this approach became widespread during the 1973 oil crisis and the 1979 energy crisis. The high cost and limited availability of fuel led to a resurgence of interest in alternatives to single occupancy vehicle travel.

Transport innovations dating from this period include high-occupancy vehicle lanes, citywide carpool systems and transportation demand management. Singapore implemented congestion pricing in the late 1970s, and Curitiba began implementing its Bus Rapid Transit system in the early 1980s.

Relatively low and stable oil prices during the 1980s and 1990s led to significant increases in vehicle travel from 1980–2000, both directly because people chose to travel by car more often and for greater distances, and indirectly because cities developed tracts of suburban housing, distant from shops and from workplaces, now referred to as urban sprawl. Trends in freight logistics, including a movement from rail and coastal shipping to road freight and a requirement for just in time deliveries, meant that freight traffic grew faster than general vehicle traffic.

At the same time, the academic foundations of the "predict and provide" approach to transport were being questioned, notably by Peter Newman in a set of comparative studies of cities and their transport systems dating from the mid-1980s. The British Government's White Paper on Transport marked a change in direction for transport planning in the UK. In the introduction to the White Paper, Prime Minister Tony Blair stated that we recognise that we cannot simply build our way out of the problems we face. It would be environmentally irresponsible-and would not work.

A companion document to the White Paper called "Smarter Choices" researched the potential to scale up the small and scattered sustainable transport initiatives then occurring across Britain, and concluded that the comprehensive application of these techniques could reduce peak period car travel in urban areas by over 20%.

A similar study by the United States Federal Highway Administration, was also released in 2004 and also concluded that a more proactive approach to transportation demand was an important component of overall national transport strategy.

Environmentally Sustainable Transport

Transport systems are major emitters of greenhouse gases, responsible for 23% of world energy-related GHG emissions in 2004, with about three quarters coming from road vehicles. Currently 95% of transport energy comes from petroleum. Energy is consumed in the manufacture as well as the use of vehicles, and is embodied in transport infrastructure including roads, bridges and railways.

The environmental impacts of transport can be reduced by improving the walking and cycling environment in cities, and by enhancing the role of public transport, especially electric rail.

Green vehicles are intended to have less environmental impact than equivalent standard vehicles, although when the environmental impact of a vehicle is assessed over the whole of its life cycle this may not be the case. Electric vehicle technology has the potential to reduce transport CO_2 emissions, depending on the embodied energy of the vehicle and the source of the electricity. Hybrid vehicles, which use an internal combustion engine combined with an electric engine to achieve better fuel efficiency than a regular combustion engine, are already common. Natural gas is also used as a transport fuel. Biofuels are a less common, and less promising, technology; Brazil met 17% of its transport fuel needs from bioethanol in 2007, but the OECD has warned that the success of biofuels in Brazil is due to specific local circumstances; internationally, biofuels are forecast to have little or no impact on greenhouse emissions, at significantly higher cost than energy efficiency measures.

In practice there is a sliding scale of green transport depending on the sustainability of the option. Green vehicles are more fuel-efficient, but only in comparison with standard vehicles, and they still contribute to traffic congestion and road crashes. Well-patronised public transport networks based on traditional diesel buses use less fuel per passenger than private vehicles, and are generally safer and use less road space than private vehicles. Green public transport vehicles including electric trains, trams and electric buses combine the advantages of green vehicles with those of sustainable transport choices. Other transport choices with very low environmental impact are cycling and other human-powered vehicles, and animal powered transport. The most common green transport choice, with the least environmental impact is walking.

Transport and Social Sustainability

Cities with overbuilt roadways have experienced unintended consequences, linked to radical drops in public transport, walking, and cycling. In many cases, streets became void of "life." Stores, schools, government centres and libraries moved away from central cities, and residents who did not flee to the suburbs experienced a much reduced quality of public space and of public services. As schools were closed their mega-school replacements in outlying areas generated additional traffic; the number of cars on US roads between 7:15 and 8:15 a.m. increases 30% during the school year.

Yet another impact was an increase in sedentary lifestyles, causing and complicating a national epidemic of obesity, and accompanying dramatically increased health care costs.

Cities and Sustainable Transport

Cities are shaped by their transport systems. In The City in History, Lewis Mumford documented how the location and layout of cities was shaped around a walkable centre, often located near a port or waterway, and with sububs accessible by animal transport or, later, by rail or tram lines.

In 1939, the New York World's Fair included a model of an imagined city, built around a car-based transport system. In this "greater and better world of tomorrow", residential, commercial and industrial areas were separated, and skyscrapers loomed over a network of urban motorways. These ideas captured the popular imagination, and are credited with influencing city planning from the 1940s to the 1970s.

The popularity of the car in the post-war era led to major changes in the structure and function of cities. There was some opposition to these changes at the time.

The writings of Jane Jacobs, in particular The Death and Life of Great American Cities provide a poignant reminder of what was lost in this transformation, and a record of community efforts to resist these changes. Lewis Mumford asked "is the city for cars or for people?" Donald Appleyard documented the consequences for communities of increasing car traffic in "The View from the Road" (1964) and in the UK, Mayer Hillman first published research into the impacts of traffic on child independent mobility in 1971. Despite these notes of caution, trends in car ownership, car use and fuel consumption continued steeply upward throughout the post-war period.

Mainstream transport planning in Europe has, by contrast, never been based on assumptions that the private car was the best or only solution for urban mobility. For example the Dutch Transport Structure Scheme has since the 1970s required that demand for additional vehicle capacity only be met "if the contribution to societal welfare is positive", and since 1990 has included an explicit target to halve the rate of growth in vehicle traffic.

Some cities outside Europe have also consistently linked transport to sustainability and to land use planning, notably Curitiba, Brazil, Portland, Oregon and Vancouver, Canada.

There are major differences in transport energy consumption between cities; an average U.S. urban dweller uses 24 times more energy annually for private transport than a Chinese urban resident, and almost four times as much as a European urban dweller. These differences cannot be explained by wealth alone but are closely linked to the rates of walking, cycling, and public transport use and to enduring features of the city including urban density and urban design. The cities and nations that have invested most heavily in car-based transport systems are now the least environmentally sustainable, as measured by per capita fossil fuel use. The social and economic sustainability of car-based urban planning has also been questioned.

Within the United States, residents of sprawling cities make more frequent and longer car trips, while residents of traditional urban neighbourhoods make a similar number of trips, but travel shorter distances and walk, cycle and use transit more often. It has been calculated that New York residents save $19 billion each year simply by owning fewer cars and driving less than the average American.

The European Commission adopted the Action Plan on urban mobility on 2009-09-30 for sustainable urban mobility. The European Commission will conduct a review of the implementation of the Action Plan in the year 2012, and will assess the need for further action.

In 2007, 72% of the European population lived in urban areas, which are key to growth and employment. Cities need efficient transport systems to support their economy and the welfare of their inhabitants. Around 85% of the EU's GDP is generated in cities. Urban areas face today the challenge of making transport sustainable in environmental (CO2, air pollution, noise) and competitiveness (congestion) terms while at the same time addressing social concerns. These range from the need to respond to health problems and demographic trends, fostering economic and social cohesion to taking into account the needs of persons with reduced mobility, families and children.

Sustainable Transport Policies and Governance

Sustainable transport policies have their greatest impact at the city level. Outside Western Europe, cities which have consistently included sustainability as a key consideration in transport and land use planning include Curitiba, Brazil, Bogota, Colombia Portland, Oregon and Vancouver, Canada. Many other cities throughout the world have recognised the need to link sustainability and transport policies, for example by joining Cities for Climate Protection.

Community and Grassroots Action

Sustainable transport is fundamentally a grassroots movement, albeit one which is now recognised as of citywide, national and international significance.

Whereas it started as a movement driven by environmental concerns, over these last years there has been increased emphasis on social equity and fairness issues, and in particular the need to ensure proper access and services for lower income groups and people with mobility limitations, including the fast growing population of older citizens.

Many of the people exposed to the most vehicle noise, pollution and safety risk have been those who do not own, or cannot drive cars, and those for whom the cost of car ownership causes a severe financial burden.

Recent Trends

Car travel increased steadily throughout the twentieth century, but trends since 2000 have been more complex. Oil price rises from 2003 have been linked to a decline in per capita fuel use for private vehicle travel in the USA, Britain and Australia. In 2008, global oil consumption fell by 0.8% overall, with significant declines in consumption in North America, Western Europe, and parts of Asia.

Criticism

The term *Green transport* is often used as a greenwash marketing technique for products which are not proven to make a positive contribution to environmental sustainability. Such claims can be legally challenged. For instance Norway's consumer ombudsman has targeted automakers who claim that their cars are "green", "clean" or "environmentally friendly". Manufacturers risk fines if they fail to drop the words. The Australian Competition and Consumer Commission (ACCC) describes *green* claims on products as *very vague, inviting consumers to give a wide range of meanings to the claim, which risks misleading them*. In 2008 the ACCC forced a car retailer to stop its *green* marketing of Saab cars, which was found by the Australian Federal Court as *misleading*.

6

Accessibility and Flow Models

Transportation is not a science, but a field of inquiry and application. As such, it tends to rely on a set of specific methodologies since transportation is a performance driven activity and this performance can be measured and compared. Transportation planning and analysis are interdisciplinary by nature, involving among others, civil engineers, economists, urban planners and geographers. Each discipline has developed methodologies dealing with their respective array of problems. Still, transportation is an infrastructure intensive activity, implying that engineering has been the dominant methodological paradigm for transportation studies. Two common traits of transportation studies, regardless of disciplinary affiliation, are a heavy reliance on empirical data and the intensive use of data analytic techniques, ranging from simple descriptive measures to more complex modeling structures.

In some respects, transport geography stands out from many other fields of human geography by the nature and function of its quantitative analysis. In fact, transport geography was one of the main forces in the quantitative revolution that helped to redefine geography in the 1960s with the use of inferential statistics, abstract models and new theories. Although this perspective provided much needed rigor, it also favoured a disconnection between empirical and theoretical approaches. Like in economics, the quantitative revolution led to a mechanistic perspective where concordance to reality became somewhat secondary; realities were made to fit into models. Even if contemporary transport geography has a more diversified approach, the quantitative dimension still plays an important part in the discipline.

Thus, in addition to providing a conceptual background to the analysis of movements of freight, people and information, transport geography is much an applied science. The main goal of methods aims to improve the efficiency of movements by identifying their spatial constraints. It is consequently possible to identify relevant strategies and policies and provide some scenarios about their possible consequences.

There are various ways of classifying the methods that are used by transport geographers:

- Whether they are qualitative or quantitative.
- Whether they deal with infrastructures (e.g. terminals) or flows.
- Whether it provides interpolation or extrapolation.
- Whether the technique provides description, explanation or optimization.
- According to the level of data aggregation, the nature of the assumptions or the complexity of the calculations.

A basic taxonomy can divide them into transport-related methods and multidisciplinary methods.

Transport-Related Methods

A first group of methods concern those directly related to the study of transportation since most draw their origins from transport planning. The methods mainly used in transport geography include:

- Network analysis (also referred to as graph theory), which is used to study transport network form and structure, especially over time. For example, one could use network analysis to study the evolution of the hub-and-spoke configuration of airline service in North America.
- Transport geographers also play a key role in studying land use-transport interactions. Numerical models have been developed, which, over time, have become increasingly complex.
- Transport geographers are also interested in flow and location allocation models that can be used to define such things as school district boundaries or the location for a new retail outlet. These techniques are optimization procedures rather than methods for describing or understanding current transport systems.

In addition, there are various methods of general use in transportation studies:

- First, a diverse set of techniques is used in the four-stage urban transportation modeling exercise, the purpose of which is to understand and predict urban spatial patterns.
- Second, traffic surveys that are used to gather empirical information about movements.

Multidisciplinary Methods

Include the whole range of methods that were not specifically developed for transportation studies, but are readily applicable to its analysis. They are labelled as multidisciplinary since they can be applied to a wide range of issues. First, there are methods that are central to geography, but are not restricted to the study of transportation systems:

- Cartography is the most obvious example of a geographic technique. Indeed, various types of maps are used in the analysis of transport systems, including land use maps, depictions of transport infrastructure, is oline maps of transportation costs, schematics of transportation activity patterns, and many more.
- Geographic information systems (GIS), which are an outgrowth of digital cartography, provide a set of tools for storing, retrieving, analysing and displaying spatial data from the real world. GIS technology has been applied to some large-scale transportation planning and engineering applications. More often, however, GIS are applied in a prescriptive way to small-scale problems, for example to plot optimal routes for buses, delivery trucks, or emergency vehicles.
- There are also various statistics that have been developed or modified by geographers to describe urban-economic systems. Examples include the Gini coefficient and indexes of concentration and specialization.

  Second, there are various methods that are used in many different applications, including transportation analysis. They underline that Transportation analysts are not restricted to those methods that have been developed with transportation in mind, but to whatever is relevant to a specific problem. In fact, many methods that were initially developed for other problems have widespread use in transportation studies:

- Some methods are used to collect primary data, e.g. questionnaires and interviews, while others are used to analyze data. Some of the analytic techniques are straightforward to implement and interpret; graphs (e.g. scattergrams, distance-decay curves) and tables (e.g. origin-destination matrices) are two examples. Others are more complex, such as inferential statistics like the t-test, analysis of variance, regression and chi-square.

- Increasingly, transportation studies are concerned with impacts and public policy issues. They relay more on qualitative information such as policy statements, rules and regulations. Various types of impacts are considered, including economic (e.g. community development), social (e.g. access to services), environmental (e.g. air or water pollution) and health (e.g. road accidents). The broad fields of environmental impact assessment, risk assessment and policy analysis are relevant to these issues.

Traffic

Traffic on roads may consist of pedestrians, ridden or herded animals, vehicles, streetcars and other conveyances, either singly or together, while using the public way for purposes of travel. Traffic laws are the laws which govern traffic and regulate vehicles, while rules of the road are both the laws and the informal rules that may have developed over time to facilitate the orderly and timely flow of traffic. Organized traffic generally has well-established priorities, lanes, right-of-way, and traffic control at intersections.

Traffic is formally organized in many jurisdictions, with marked lanes, junctions, intersections, interchanges, traffic signals, or signs. Traffic is often classified by type: heavy motor vehicle (e.g., car, truck); other vehicle (e.g., moped, bicycle); and pedestrian. Different classes may share speed limits and easement, or may be segregated. Some jurisdictions may have very detailed and complex rules of the road while others rely more on drivers' common sense and willingness to cooperate. Organization typically produces a better combination of travel safety and efficiency. Events which disrupt the flow and may cause traffic to degenerate into a disorganized mess include: road construction, collisions and debris in the roadway. On particularly busy freeways, a minor disruption may persist in a phenomenon known as traffic waves. A complete breakdown of organization may result in traffic jams and gridlock. Simulations of organized traffic frequently involve queuing theory, stochastic processes and equations of mathematical physics applied to traffic flow.

Rules of the Road

Rules of the road are the general practices and procedures that road users are required to follow. These rules usually apply to all road users, though they are of special importance to motorists and cyclists. These rules govern interactions between vehicles and with pedestrians. The basic traffic rules are defined by an international treaty under the authority of the United Nations, the 1968 *Vienna Convention on Road Traffic*. Not all countries are signatory to the convention and, even among signatories, local variations in practice may be found. There are also unwritten local rules of the road, which are generally understood by local drivers.

As a general rule, drivers are expected to avoid a collision with another vehicle and pedestrians, regardless of whether or not the applicable rules of the road allow them to be where they happen to be. In addition to the rules applicable by default, traffic signs and traffic lights must be obeyed, and instructions may be given by a police officer, either routinely (on a busy crossing instead of traffic lights) or as road traffic control around a construction zone, accident, or other road disruption. These rules should be distinguished from the mechanical procedures required to operate one's vehicle.

Directionality

Traffic going in opposite directions should be separated in such a way that they do not block each other's way. The most basic rule is whether to use the left or right side of the road.

Traffic Regulations

In many countries, the rules of the road are codified, setting out the legal requirements and punishments for breaking them. In the United Kingdom, the rules are set out in the Highway Code, which includes obligations but also advice on how to drive sensibly and safely.

In the United States, traffic laws are regulated by the states and municipalities through their respective traffic codes. Most of these are based at least in part on the Uniform Vehicle Code, but there are variations from state to state. In states such as Florida, traffic law and criminal law are separate, therefore, unless someone flees a scene of an accident, commits vehicular homicide or manslaughter, they are only guilty of a minor traffic offense. However, states such as South Carolina have completely criminalized their traffic law, so, for example, you are guilty of a misdemeanour simply for travelling 5 miles over the speed limit.

Organized Traffic

Priority (Right of Way)

Vehicles often come into conflict with other vehicles and pedestrians because their intended courses of travel intersect, and thus interfere with each other's routes. The general principle that establishes who has the right to go first is called "right of way", or "priority". It establishes who has the right to use the conflicting part of the road and who has to wait until the other does so. Signs, signals, markings and other features are often used to make priority explicit. Some signs, such as the stop sign, are nearly universal. When there are no signs or markings, different rules are observed depending on the location. These default priority rules differ between countries, and may even vary within countries. Trends toward uniformity are exemplified at an international level by the Vienna Convention on Road Signs and Signals, which prescribes standardized traffic control devices (signs, signals, and markings) for establishing the right of way where necessary.

Crosswalks (or pedestrian crossings) are common in populated areas, and may indicate that pedestrians have priority over vehicular traffic. In most modern cities, the traffic signal is used to establish the right of way on the busy roads. Its primary purpose is to give each road a duration of time in which its traffic may use the intersection in an organized way. The intervals of time assigned for each road may be adjusted to take into account factors such as difference in volume of traffic, the needs of pedestrians, or other traffic signals. Pedestrian crossings may be located near other traffic control devices; if they are not also regulated in some way, vehicles must give priority to them when in use. Traffic on a public road usually has priority over other traffic such as traffic emerging from private access; rail crossings and drawbridges are typical exceptions.

Uncontrolled Traffic

Uncontrolled traffic occurs in the absence of lane markings and traffic control signals. On roads without marked lanes, drivers tend to keep to the appropriate side if the road is wide enough. Drivers frequently overtake others. Obstructions are common.

Intersections have no signals or signage, and a particular road at a busy intersection may be dominant – that is, its traffic flows – until a break in traffic, at which time the dominance shifts to the other road where vehicles are queued. At the intersection of two perpendicular roads, a traffic jam may result if four vehicles face each other side-on.

Turning

Drivers will often want to cease to travel a straight line and turn onto another road or onto private property. The vehicle's directional signals (blinkers) are often used as a way to announce one's intention to turn, thus alerting other drivers. The actual usage of blinkers varies greatly amongst countries, although its purpose should be the same in all countries: to indicate a driver's intention to depart from the current (and natural) flow of traffic well before the departure is executed (typically 3 seconds as a guideline).

This will usually mean that turning traffic will have to stop in order to wait for a breach to turn, and this might cause inconvenience for drivers that follow them but do not want to turn. This is why dedicated lanes and protected traffic signals for turning are sometimes provided. On busier intersections where a protected lane would be ineffective or cannot be built, turning may be entirely prohibited, and drivers will be required to "drive around the block" in order to accomplish the turn. Many cities employ this tactic quite often; in San Francisco, due to its common practice, making three right turns is known colloquially as a "San Francisco left turn". Likewise, as many intersections in Taipei City are too busy to allow direct left turns, signs often direct drivers to drive around the block to turn.

Turning rules are by no means universal. In New Zealand, for example, left turning traffic must give way to opposing "right turning" traffic, i.e., traffic turning into a driver's path (unless there are multiple lanes to turn into).

On roads with multiple lanes, turning traffic is generally expected to move to the lane closest to the direction they wish to turn. For example, traffic intending to turn right will usually move to the rightmost lane before the intersection. Likewise, left-turning two rightmost lanes will be of authority; for example, in Brazil and elsewhere it is common for drivers to observe (and trust) the turn signals used by other drivers in order to make turns from other lanes. For example if several vehicles on the right lane are all turning right, a vehicle may come from the next-to-right lane and turn right as well, doing so in parallel with the other right-turning vehicles.

Intersections

In most of Continental Europe, the default rule is to give priority to the right, but this may be overridden by signs or road markings, and does not apply at T-shaped junctions in some of these countries, such as France. There, priority was initially given according to the social rank of each traveller, but early in the life of the automobile this rule was deemed impractical and replaced with the priority a droite (priority to the right) rule, which still applies.

At a traffic circle where priority a droite is not overridden, traffic on what would otherwise be a roundabout gives way to traffic entering the circle. Most French roundabouts now have give-way signs for traffic entering the circle, but there remain some notable exceptions that operate on the old rule, such as the Place de l'Etoile around the Arc de Triomphe. Traffic at this intersection is so chaotic that French insurance companies deem any accident on the roundabout to be equal liability. Priority to the right where used in continental Europe may be overridden by an ascending hierarchy of markings, signs, signals, and authorized persons.

In the United Kingdom, priority is always indicated by signs or markings, so that every junction between public roads (except those governed by traffic signals) has a concept of a major road and minor road. The default give-way-to-the-right rule used in Continental Europe causes problems for many British and Irish drivers who are accustomed to having right of way by default unless they are specifically told to give way.

Other countries use various methods similar to the above examples to establish the right of way at intersections. For example, in most of the United States, the default priority is to yield to traffic from the right, but this is usually overridden by traffic control devices or other rules, like the boulevard rule. This rule holds that traffic entering a major road from a smaller road or alley must yield to the traffic of the busier road, but signs are often still posted.

The boulevard rule can be compared with the above concept of a major and minor road, or the priority roads that may be found in countries that are parties to the Vienna Convention on Road Signs and Signals.

Perpendicular intersections Also known as a "four-way" intersection, this intersection is the most common configuration for roads that cross each other, and the most basic type.

If traffic signals do not control a 4-way intersection, signs or other features are typically used to control movements and make clear priorities. The most common arrangement is to indicate that one road has priority over the other, but there are complex cases where all traffic approaching an intersection must yield and may be required to stop.

In the United States, South Africa, and Canada, there are four-way intersections with a stop sign at every entrance, called four-way stops. A failed signal or a flashing red light is equivalent to a four-way stop, or an all-way stop. Special rules for four-way stops may include:

1. In the countries that use four-way stops, pedestrians always have priority at crosswalks – even at unmarked ones, which exist as the logical continuations of the sidewalks at every intersection with approximately right angles – unless signed or painted otherwise.
2. Whichever vehicle first stops at the stop line – or before the crosswalk, if there is no stop line – has priority.
3. If two vehicles stop at the same time, priority is given to the vehicle on the right.
4. If three vehicles stop at the same time, priority is given to the two vehicles going in opposite directions, if possible.
5. If four vehicles stop, drivers usually use gestures and other communication to establish right-of-way.

In Europe and other places, there are similar intersections. These may be marked by special signs (according to the Vienna Convention on Road Signs and Signals), a danger sign with a black X representing a crossroads. This sign informs drivers that the intersection is uncontrolled and that default rules apply. In Europe and in many areas of North America the default rules that apply at uncontrolled four-way intersections are almost identical:

1. Rules for pedestrians differ by country, in the United States and Canada pedestrians generally have priority at such an intersection.
2. All vehicles must give priority to any traffic approaching from their right,
3. Then, if the vehicle is turning right or continuing on the same road it may proceed.
4. Vehicles turning left must also give priority to traffic approaching from the opposite direction, unless that traffic is also turning left.
5. If the intersection is congested, vehicles must alternate directions and/or circulate priority to the right one vehicle at a time.

Pedestrian Crossings

Pedestrians must often cross from one side of a road to the other, and in doing so may come into the way of vehicles travelling on the road. In many places pedestrians are entirely left to look after themselves, that is, they must observe the road and cross when they can see that no traffic will threaten them. Busier cities usually provide pedestrian crossings, which are strips of the road where pedestrians are expected to cross. The actual appearance of pedestrian crossings varies greatly, but the two most common appearances are: (1) a series of parallel white stripes or (2) two long horizontal white lines. The former is usually preferred, as it stands out more conspicuously against the dark pavement.

Some pedestrian crossings also accompany a traffic signal which will make vehicles stop at regular intervals so the pedestrians can cross. Some countries have "intelligent" pedestrian signals, where the pedestrian must push a button in order to assert his intention to cross. The traffic signal will use that information to schedule itself, that is, when no pedestrians are present the signal will never pointlessly cause vehicle traffic to stop.

Pedestrian crossings without traffic signals are also common. In this case, the traffic laws usually states that the pedestrian has the right of way when crossing, and that vehicles must stop when a pedestrian uses the crossing. Countries and driving cultures vary greatly as to the extent to which this is respected. In the state of Nevada the car has the right of way when the crosswalk signal specifically forbids pedestrian crossing. Some jurisdictions forbid crossing or using the road anywhere other than at crossings, termed jaywalking. In other areas, pedestrians may have the right to cross where they choose, and have right of way over vehicular traffic while crossing. In most areas, an intersection is considered to have a crosswalk, even if not painted, as long as the roads meet at approximate right angles. Examples of locations where this rule is not in effect are the United Kingdom and Croatia.

Level Crossings

A level crossing is an at-grade intersection of a railway by a road. Because of safety issues, they are often equipped with closable gates, crossing bells and warning signs.

Speed Limits

The higher the speed of a vehicle, the more difficult collision avoidance becomes and the greater the damage if a collision does occur. Therefore, many countries of the world limit the maximum speed allowed on their roads. Vehicles are not supposed to be driven at speeds which are higher than the posted maximum.

To enforce speed limits, two approaches are generally employed. In the United States, it is common for the police to patrol the streets and use special equipment (typically a radar unit) to measure the speed of vehicles, and pull over any vehicle found to be in violation of the speed limit. In Brazil and some European countries, there are computerized speed-measuring devices spread throughout the city, which will automatically detect speeding drivers and take a photograph of the license plate (or number plate), which is later used for applying and mailing the ticket. Many jurisdictions in the U.S. use this technology as well.

A mechanism that was developed in Germany is the Grune Welle, or green wave, which is an indicator that shows the optimal speed to travel for the synchronized green lights along that corridor. Driving faster or slower than the speed set by the behaviour of the lights causes the driver to frequently encounter red lights. This discourages drivers from speeding or impeding the flow of traffic.

Overtaking

Overtaking (or passing) refers to a manoeuvre by which one or more vehicles travelling in the same direction are passed by another vehicle. On two-lane roads, when there is a split line or a dashed line on the side of the overtaker, drivers may overtake when it is safe. On multi-lane roads in most jurisdictions, overtaking is permitted in the "slower" lanes, though many require a special circumstance.

In the United Kingdom and Canada, notably on extra-urban roads, a solid white or yellow line closer to the driver is used to indicate that no overtaking is allowed in that lane. A double white or yellow line means that neither side may overtake.

In the United States, a solid white line means that lane changes are discouraged and a double-white line means that the lane change is prohibited.

Lanes

When a street is wide enough to accommodate several vehicles travelling side-by-side, it is usual for traffic to organize itself into lanes, that is, parallel corridors of traffic. Some roads have one lane for each direction of travel and others have multiple lanes for each direction. Most countries apply pavement markings to clearly indicate the limits of each lane and the direction of travel that it must be used for. In other countries lanes have no markings at all and drivers follow them mostly by intuition rather than visual stimulus.

On roads that have multiple lanes going in the same direction, drivers may usually shift amongst lanes as they please, but they must do so in a way that does not cause inconvenience to other drivers. Driving cultures vary greatly on the issue of "lane ownership": in some countries, drivers travelling in a lane will be very protective of their right to travel in it while in others drivers will routinely expect other drivers to shift back and forth.

Designation and Overtaking

The usual designation for lanes on divided highways is the fastest lane is the one closest to the centre of the road, and the slowest to the edge of the road. Drivers are usually expected to keep in the slowest lane unless overtaking, though with more traffic congestion all lanes are often used.

When driving on the left:

- The lane designated for faster traffic is on the right.
- The lane designated for slower traffic is on the left.
- Most freeway exits are on the left.
- Overtaking is permitted to the right, and sometimes to the left.

When driving on the right:

- The lane designated for faster traffic is on the left.
- The lane designated for slower traffic is on the right.
- Most freeway exits are on the right.
- Overtaking is permitted to the left, and sometimes to the right.

Countries party to the Vienna Convention on Road Traffic have uniform rules about overtaking and lane designation. The convention details (amongst other things) that "Every driver shall keep to the edge of the carriageway appropriate to the direction of traffic", and the "Drivers overtaking shall do so on the side opposite to that appropriate to the direction of traffic", notwithstanding the presence or absence of oncoming traffic. Allowed exceptions to these rules include turning or heavy traffic, traffic in lines, or situation in which signs or markings must dictate otherwise.

These rules must be more strictly adhered to on roads with oncoming traffic, but still apply on multi-lane and divided highways. Many countries in Europe are party to the Vienna Conventions on traffic and roads. In Australia (which is not a contracting party), travelling in any lane other than the "slow" lane with a speed limit at or above 80 km/h (50 mph) is an offence, unless signage is posted to the contrary or the driver is overtaking.

Many areas in North America do not have any laws about staying to the slowest lanes unless overtaking. In those areas, unlike many parts of Europe, traffic is allowed to overtake on any side, even in a slower lane. This practice is known as "passing on the right" in the United States (where it is common) and "overtaking on the inside" and "undertaking" in the United Kingdom. In most countries, the inside lane refers to the fastest lane (the lane closest to the highway median), but in the United Kingdom, it refers to the slowest lane (the lane that is in fact outside).

U.S.-State-specific Practices

In some U.S. states (such as Louisiana, Massachusetts and New York), although there are laws requiring all traffic on a public way to use the right-most lane unless overtaking, this rule is often ignored and seldom enforced on multi-lane roadways. Some states, such as Colorado, use a combination of laws and signs restricting speeds or vehicles on certain lanes to emphasize overtaking only on the left lane, and to avoid a psychological condition commonly called road rage.

In California, cars may use any lane on multi-lane roadways. Drivers moving slower than the general flow of traffic are required to stay in the right-most lanes (by California Vehicle Code (CVC) 21654) to keep the way clear for faster vehicles and thus speed up traffic. However, faster drivers may legally pass in the slower lanes if conditions allow (by CVC 21754). But the CVC also requires trucks to stay in the right lane, or in the right two lanes if the roadway has four or more lanes going in their direction. The oldest freeways in California, and some freeway interchanges, often have ramps on the left, making signs like "Trucks ok on left Lane" or "Trucks may use all Lanes" necessary to override the default rule. Lane splitting, or riding motorcycles in the space between cars in traffic, is permitted as long as it is done in a safe and prudent manner.

One-way Roadways

In order to increase traffic capacity and safety, a route may have two or more separate roads for each direction of traffic. Alternatively, a given road might be declared one-way.

Expressways and Freeways

In large cities, moving from one part of the city to another by means of ordinary streets and avenues can be time-consuming since traffic is often slowed by at-grade junctions, tight turns, narrow marked lanes and lack of a minimum speed limit. Therefore, it has become common practice for larger cities to build expressways or freeways, which are large and wide roadways with limited access, that typically run for long distances without at-grade junctions.

The words expressway and freeway have varying meanings in different jurisdictions and in popular use in different places; however, there are two different types of roads used to provide high-speed access across urban areas:

- The freeway (in U.S. usage) or motorway in UK usage, is a divided multi-lane highway with fully-controlled access and grade-separated intersections (no cross traffic). Some freeways are called expressways, super-highways, or turnpikes, depending on local usage. Access to freeways is fully controlled; entering and leaving the freeway is permitted only at grade-separated interchanges.
- The expressway (when the name does not refer to a freeway or motorway) is usually a broad multi-lane avenue, frequently divided, with some grade-level intersections (although usually only where other expressways or arterial roads cross).

Motor vehicle drivers wishing to travel over great distances within the city will usually take the freeways or expressways in order to minimize travel time. When a crossing road is at the same grade as the freeway, a bridge (or, less often, an underpass) will be built for the crossing road. If the freeway is elevated, the crossing road will pass underneath it.

Minimum speed signs are sometimes posted (although increasingly rare) and usually indicate that any vehicle travelling slower than 40 mph (64 km/h) should indicate a slower speed of travel to other motor vehicles by engaging the vehicle's four-way flashing lights. Alternative slower-than-posted speeds may be in effect, based on the posted speed limit of the highway/freeway.

Systems of freeways and expressways are also built to connect distant and regional cities, notable systems include the Interstate highways, the Autobahnen and the Expressway Network of the People's Republic of China.

One-way Streets

In more sophisticated systems such as large cities, this concept is further extended: some streets are marked as being one-way, and on those streets all traffic must flow in only one direction, but pedestrians on the sidewalks are generally not limited to one-way movement. A driver wishing to reach a destination he already passed must use other streets in order to return. Usage of one-way streets, despite the inconveniences it can bring to individual drivers, can greatly improve traffic flow since they usually allow traffic to move faster and tend to simplify intersections.

Congested Traffic

In some places traffic volume is consistently, extremely large, either during periods of time referred to as *rush hour* or perpetually. Exceptionally, traffic upstream of an accident or an obstruction, such as construction, may also be constrained, resulting in a traffic jam. Such dynamics in relation to traffic congestion is known as traffic flow. Traffic engineers sometimes gauge the quality of traffic flow in terms of level of service.

Rush Hour

During business days in most major cities, traffic congestion reaches great intensity at predictable times of the day due to the large number of vehicles using the road at the same time. This phenomenon is called rush hour or peak hour, although the period of high traffic intensity often exceeds one hour.

Congestion Mitigation

Rush Hour Policies

Some cities adopt policies to reduce rush-hour traffic and pollution and encourage the use of public transportation. For example, in Sao Paulo, Manila and in Mexico City, each vehicle has a specific day of the week in which it is forbidden from travelling the roads during rush hour. The day for each vehicle is taken from the license plate number, and this rule is enforced by traffic police and also by hundreds of strategically positioned traffic cameras backed by computerized image-recognition systems that issue tickets to offending drivers.

In the United States and Canada, several expressways have a special lane (called an "HOV Lane"-High Occupancy Vehicle Lane) that can only be used by cars carrying two (some locations-three) or more people. Also, many major cities have instituted strict parking prohibitions during rush hour on major arterial streets leading to and from the central business district. During designated weekday hours, vehicles parked on these primary routes are subject to prompt ticketing and towing at owner expense. The purpose of these restrictions is to make available an additional traffic lane in order to maximize available traffic capacity. Additionally, several cities offer a public telephone service where citizens can arrange rides with others depending on where they live and work. The purpose of these policies is to reduce the number of vehicles on the roads and thus reduce rush-hour traffic intensity.

Metered freeways are also a solution for controlling rush hour traffic. In Phoenix, Arizona metered on-ramps have been implemented. During rush hour, traffic signals are used with green lights to allow one car per blink of the light to proceed on to the freeway.

Pre-emption

In some areas, emergency responders are provided with specialized equipment which allows emergency response vehicles, particularly fire fighting apparatus, to have high-priority travel by having the lights along their route change to green. The technology behind these methods have evolved, from panels at the fire department (which could trigger and control green lights for certain major corridors) to optical systems (which the individual fire apparatus can be equipped with to communicate directly with receivers on the signal head). In other areas, public transport buses have special equipment to get green lights.

During emergencies where evacuation of a heavily populated area is required, local authorities may institute contraflow lane reversal, in which all lanes of a road lead away from a danger zone regardless of their original flow. Aside from emergencies, contraflow may also be used to ease traffic congestion during rush hour or at the end of a sports event (where a large number of cars are leaving the venue at the same time). For example, the six lanes of the Lincoln Tunnel can be changed from three in-bound and three out-bound to a two/four configuration depending on traffic volume. The Brazilian highways Rodovia dos Imigrantes and Rodovia Anchieta connect Sao Paulo to the Atlantic coast. Almost all lanes of both highways are usually reversed during weekends to allow for heavy seaside traffic. The reversibility of the highways requires many additional highway ramps and complicated interchanges.

Intelligent Transportation Systems

An intelligent transportation system (ITS) is a system of hardware, software, and operators that allow better monitoring and control of traffic in order to optimize traffic flow. As the number of vehicle lane miles traveled per year continues to increase dramatically, and as the number of vehicle lane miles constructed per year has not been keeping pace, this has led to ever-increasing traffic congestion. As a cost-effective solution toward optimizing traffic, ITS presents a number of technologies to reduce congestion by monitoring traffic flows through the use of sensors and live cameras or analysing cellular phone data travelling in cars (floating car data) and in turn rerouting traffic as needed through the use of variable message boards (VMS), highway advisory radio, on board or off board navigation devices and other systems through integration of traffic data with navigation systems. Additionally, the roadway network has been increasingly fitted with additional communications and control infrastructure to allow traffic operations personnel to monitor weather conditions, for dispatching maintenance crews to perform snow or ice removal, as well as intelligent systems such as automated bridge de-icing systems which help to prevent accidents.

Transport Engineering

Transportation engineering is the application of scientific principles to the safe and efficient movement of people and goods (transport). It is a sub-discipline of civil engineering. Transportation engineering is a major component of the civil engineering discipline. The importance of transportation engineering within the civil engineering profession can be judged by the number of divisions in ASCE (American Society of Civil Engineers) that are directly related to transportation. There are six such divisions (Aerospace; Air Transportation; Highway; Pipeline; Waterway, Port, Coastal and Ocean; and Urban Transportation) representing one-third of the total 18 technical divisions within the ASCE (1987).

The planning aspects of transport engineering relate to urban planning, and involve technical forecasting decisions and political factors. Technical forecasting of passenger travel usually involves an urban transportation planning model, requiring the estimation of trip generation (how many trips for what purpose), trip distribution (destination choice, where is the traveller going), mode choice (what mode is being taken), and route assignment (which streets or routes are being used). More sophisticated forecasting can include other aspects of traveller decisions, including auto ownership, trip chaining (the decision to link individual trips together in a tour) and the choice of residential or business location (known as land use forecasting).

Passenger trips are the focus of transport engineering because they often represent the peak of demand on any transportation system. A review of descriptions of the scope of various committees indicates that while facility planning and design continue to be the core of the transportation engineering field, such areas as operations planning, logistics, network analysis, financing, and policy analysis are also important to civil engineers, particularly to those working in highway and urban transportation. Transportation engineering, as practiced by civil engineers, primarily involves planning, design, construction, maintenance, and operation of transportation facilities. The facilities support air, highway, railroad, pipeline, water, and even space transportation. The design aspects of transport engineering include the sizing of transportation facilities (how many lanes or how much capacity the facility has), determining the materials and thickness used in pavement designing the geometry (vertical and horizontal alignment) of the roadway (or track).

Operations and management involve traffic engineering, so that vehicles move smoothly on the road or track. Older techniques include signs, signals, markings, and tolling. Newer technologies involve intelligent transportation systems, including advanced traveller information systems (such as variable message signs), advanced traffic control systems (such as ramp meters), and vehicle infrastructure integration. Human factors are an aspect of transport engineering, particularly concerning driver-vehicle interface and user interface of road signs, signals, and markings.

Highway Engineering

Engineers in this specialization:

- Handle the planning, design, construction, and operation of highways, roads, and other vehicular facilities as well as their related bicycle and pedestrian realms.
- Estimate the transportation needs of the public and then secure the funding for the project.
- Analyze locations of high traffic volumes and high collisions for safety and capacity.
- Use civil engineering principles to improve the transportation system.

Railroad Engineering

Railway engineers handle the design, construction, and operation of railroads and mass transit systems that use a fixed guideway (such as light rail or even monorails). Typical tasks would include determining horizontal and vertical alignment design, station location and design, and construction cost estimating. Railroad engineers can also move into the specialized field of train dispatching which focuses on train movement control. Railway engineers also work to build a cleaner and safer transportation network by reinvesting and revitalizing the rail system to meet future demands. In the United States, railway engineers work with elected officials in Washington, D.C. on rail transportation issues to make sure that the rail system meets the country's transportation needs.

Port and Harbour Engineering

Port and harbour engineers handle the design, construction, and operation of ports, harbours, canals, and other maritime facilities. This is not to be confused with marine engineering.

Airport Engineering

Airport engineers design and construct airports. Airport engineers must account for the impacts and demands of aircraft in their design of airport facilities. One such example is the analysis of predominant wind direction to determine runway orientation.

Cargo

Cargo (or freight) is goods or produce transported, generally for commercial gain, by ship, aircraft, train, van or truck. In modern times, containers are used in most intermodal long-haul cargo transport.

Marine Cargo Types

There is a wide range of marine cargoes at seaport terminals operated. The primary types are these:

- Automobiles are handled at many ports, usually carried on specialist roll-on/roll-off ships.
- Break bulk cargo is typically material stacked on wooden pallets and lifted into and out of the hold of a vessel by cranes on the dock or aboard the ship itself. The volume of break bulk cargo has declined dramatically worldwide as containerization has grown. A safe and secure way to secure Break bulk and freight in containers is by using Dunnage Bags.
- Bulk cargoes, such as salt, oil, tallow, and Scrap metal, are usually defined as commodities that are neither on pallets nor in containers, and which are not handled as individual pieces, the way heavy-lift and project cargoes are. Alumina, grain,

gypsum, logs and wood chips, for instance, are bulk cargoes.

- Containers are the largest and fastest growing cargo category at most ports worldwide. Containerized cargo includes everything from auto parts and machinery components to shoes, toys, and frozen meat and seafood.
- Project cargo and heavy lift cargo may include items such as manufacturing equipment, factory components, power equipment such as generators and wind turbines, military equipment or almost any other over sized or overweight cargo too big or too heavy to fit into a container.

Air Cargo

Air cargo is commonly known as air freight. There are many firms which collect freight from a shipper and deliver it to the customer. Aircraft were first used for carrying mail as cargo in 1911, but eventually manufacturers started designing aircraft for freight as well. There are many commercial aircraft suitable for carrying cargo such as the Boeing 747 and the bigger An-124, which were purpose built to be easily converted to a cargo aircraft. Such very large aircraft also employ quick loading containers known as ULDs much like containerized cargo ships. It is located in front mismo of the aircraft the triangular shaped in front.

Most nations own and utilize large numbers of cargo aircraft such as the C-17 Globemaster III, for airlift logistics needs of such operations.

Freight Train

Trains are capable of transporting large numbers of containers which have come off the shipping ports. Trains are also used for the transportation of steel, wood and coal. Trains are used as they can pull a large amount and generally have a direct route to the destination. Under the right circumstances, freight transport by rail is more economic and energy efficient than by road, especially when carried in bulk or over long distances. The main disadvantage of rail freight is its lack of flexibility. For this reason, rail has lost much of the freight business to road transport. Rail freight is often subject to transshipment costs since it must be transferred from one mode to another in the chain; these costs may dominate and practices such as containerization aim at minimizing these. Many governments are now trying to encourage more freight onto trains, because of the environmental benefits that it would bring; rail transport is very energy efficient.

Van or Truck Cargo

There are many firms which transport all types of cargo, ranging from letters to houses to cargo containers. These firms like Parcelforce or FedEx which deliver fast and sometimes same day deliverly services. A good example of road cargo is supermarket stock, as these require deliveries every day to keep the shelves stacked with goods for sale. Retailers of all kinds rely upon delivery trucks, be they full size semi trucks or smaller delivery vans. Freight is a term used to classify the transportation of cargo and is typically a commercial process. Items are usually organized into various shipment categories before they are transported. This is dependent on several factors:

- The type of item being carried, i.e. a kettle could fit into the category 'household goods'.
- How large the shipment is, both in terms of item size and quantity.
- How long the item for delivery will be in transit.

Shipments are typically categorized as household goods, express, parcel, and freight shipments.

Furniture, art, or similar items are usually classified as "household goods" (HHG).

Very small business or personal items like envelopes are considered "overnight express" or "express letter" shipments. These shipments are rarely over a few kilos/pounds, and almost always travel in the carrier's own packaging. Service levels are variable, depending on the shipper's choice. Express shipments almost always travel some distance by air. An envelope may go USA coast to USA coast overnight or it may take several days, depending on the service options and prices chosen.

Larger items like small boxes are considered "parcel" or "ground" shipments. These shipments are rarely over 50 kg (110 lb), with no single piece of the shipment weighing more than about 70 kg (154 lb). Parcel shipments are always boxed, sometimes in the shipper's packaging and sometimes in carrier-provided packaging. Service levels are again variable; but most "ground" shipments will move about 800 to 1,100 kilometres (497 to 684 mi) per day, going coast to coast in about four days depending on origin. Parcel shipments rarely travel by air, and typically move via road and rail. Parcels represent the majority of business-to-consumer (B2C) shipments.

Beyond HHG, express, and parcel shipments, movements are termed "freight shipments."

Less-than-truckload Freight

"Less than truckload" (LTL) cargo is the first category of freight shipment, and represents the majority of "freight" shipments and the majority of business-to-business (B2B) shipments. LTL shipments are also often referred to as "motor freight" and the carriers involved are referred to as "motor carriers". LTL shipments range from 50 to 7,000 kg (110 to 15,000 lb), and the majority of times they will be less than 2.5 to 8.5 m (8 ft 2.4 in to 27 ft 10.6 in). The average single piece of LTL freight is 600 kg (1,323 lb) and the size of a standard pallet. Long freight and/or large freight are subject to "extreme length" and "cubic capacity" surcharges.

Trailers used in LTL can range from 28 to 53 ft (8.53 to 16.15 m). The standard for city deliveries is usually 48 ft (14.63 m). In tight and residential environments the 28 ft (8.53 m) trailer is used the most. The shipments are usually palletized, shrink-wrapped and packaged for a mixed-freight environment. Unlike express or parcel, LTL shippers must provide their own packaging, as LTL carriers do not provide any packaging supplies or assistance. However, crating or other substantial packaging may be required for LTL shipments in circumstances that require this criteria.

"Air cargo" or "air freight" shipments are very similar to LTL shipments in terms of size and packaging requirements. However, air freight shipments typically need to move at much faster speeds than 800 km or 497 mi per day. Air shipments may be booked directly with the carriers or through brokers or online marketplace services. While shipments move faster than standard LTL, "air" shipments don't always actually move by air.

Truckload Freight

In the United States of America, shipments larger than about 7,000 kg (15,432 lb) are typically classified as "truckload" (TL), given that it is more efficient and economical for a large shipment to have exclusive use of one larger trailer rather than share space on a smaller LTL trailer. The total weight of a loaded truck (tractor and trailer, 5-axle rig) cannot exceed 36,000 kg (79,366 lb) in the U.S. In ordinary circumstances, long-haul equipment will weigh about 15,000 kg (33,069 lb); leaving about 20,000 kg (44,092 lb) of freight capacity. Similarly a load is limited to the space available in the trailer; normally

48 ft (14.63 m) or 53 ft (16.15 m) long and 2.6 m (102.4 in) wide and 2.7 m (8 ft 10.3 in) high (13 ft 6 in/4.11 m high over all). While express, parcel, and LTL shipments are always intermingled with other shipments on a single piece of equipment and are typically reloaded across multiple pieces of equipment during their transport, TL shipments usually travel as the only shipment on a trailer and TL shipments usually deliver on exactly the same trailer as they are picked up on.

Often, an LTL shipper may realize savings by utilizing a freight "broker," online marketplace, or other intermediary instead of contracting directly with a trucking company. Brokers can shop the marketplace and obtain lower rates than most smaller shippers can directly. In the Less-than-Truckload (LTL) marketplace, intermediaries typically receive 50% to 80% discounts from published rates, where a small shipper may only be offered a 5% to 30% discount by the carrier. Intermediaries are licensed by the DOT and have requirements to provide proof of insurance.

Truckload (TL) carriers usually charge a rate per kilometre or mile. The rate varies depending on the distance, geographic location of the delivery, items being shipped, equipment type required, and service times required. TL shipments usually receive a variety of surcharges very similar to those described for LTL shipments above. In the TL market, there are thousands more small carriers than in the LTL market; so the use of transportation intermediaries or "brokers" is extremely common.

Another cost-saving method is facilitating pickups or deliveries at the carrier's terminals. By doing this, shippers avoid any accessorial fees that might normally be charged for liftgate, residential pickup/delivery, inside pickup/delivery or notifications/appointments. Carriers or intermediaries can provide shippers with the address and phone number for the closest shipping terminal to the origin and/or destination.

Shipping experts optimize their service and costs by sampling rates from several carriers, brokers, and online marketplaces. When obtaining rates from different providers, shippers may find quite a wide range in the pricing offered. If a shipper uses a broker, freight forwarder, or other transportation intermediary, it is common for the shipper to receive a copy of the carrier's Federal Operating Authority. Freight brokers and intermediaries are also required by Federal Law to be licensed by the Federal Highway Administration. Experienced shippers avoid unlicensed brokers and forwarders; because if brokers are working outside the law by not having a Federal Operating License, the shipper has no protection in the event of a problem. Also shippers normally ask for a copy of the broker's insurance certificate and any specific insurance that applies to the shipment.

United States Security Concerns

Cargo represents a concern to U.S. national security. It was reported from Washington, DC in 2003 that over 6 million cargo containers enter the United States ports each year. After the terrorist attacks of September 11th, the security of this magnitude of cargo has become highlighted. The latest US Government response to this threat is the CSI: Container Security Initiative. CSI is a program intended to help increase security for containerized cargo shipped to the United States from around the world.

Load Securing

There are many different ways and materials available to stabilize and secure cargo in the various modes of transportation. Conventional load securing methods and materials such as steel strapping and wood blocking & bracing have been used for decades and are still widely used. Present load securing methods offer several other options including polyester strapping and lashing, synthetic webbings and dunnage bags, also known as air bags or inflatable bags.

Urban Transport

Over the last two decades, rapid population growth and spatial expansion has led to a sharp increase in demand for urban transport facilities and services in many cities in the ESCAP region. However, several factors have hindered the adequate provision of services to match the ever-increasing demand. In many cities, densification and spatial expansion have occurred with little or no development planning, while in some cases the failure of the instruments of governance has resulted in a significant wastage of resources or substandard quality of infrastructure. Furthermore, the huge capital costs and time required to develop high capacity transit systems have prevented the timely implementation of such systems in rapidly growing urban areas. As a result, many cities have relied on road-based systems, which have serious capacity constraints, negative environmental consequences and other limitations.

Consequently, many cities in the region are facing serious problems, including significant levels of traffic congestion, air pollution from transport sources, high rates of traffic accidents and inadequate access to transport facilities by poor and vulnerable groups, such as people with disabilities. The deteriorating urban environment threatens the "liveability" and productivity of many cities. In some of the major capitals, such as Bangkok,

Dhaka, Manila and New Delhi, the situation is so severe that the efficiency of their urban economy is negatively affected, as is the health and welfare of the people living in them.

Growing Motorization

While the level of motorization in Asian cities is still much lower than levels in European cities, a trend of rapid motorization is evident in almost all of them. Apart from a few cities in Central Asia, there has been a considerable increase in the motor vehicle populations of all major cities. However, there are significant variations in the level and rate of motorization between cities, due partly to differences in income levels and government policy.

For example, the number of road vehicles in Bangkok grew from 1 million in 1984 to 4 million in 1998.113 The vehicle population of Beijing grew about threefold between 1991 and 2000, from 540,000 to 1,570,000. Similar trends in growth occurred in Jakarta and Kuala Lumpur. Since the late 1990s, Indian cities have also experienced rapid growth rates of their vehicular population, following the introduction of economic reforms that lowered costs and increased the affordability of passenger cars. For example, Mumbai has registered an annual growth of motorized vehicles of about 10 per cent in recent years, while between 1995 and 2000, Delhi's total motor vehicle population grew from 2.4 to 3.3 million, of which the car population increased from 576,000 to 837,000.114

The exponential growth of motorized two-and three-wheelers is another visible trend in Asian cities. In many countries, such as Thailand, Malaysia, and Indonesia, two-and threewheelers make up over half of all motor vehicles. Bangkok currently has an estimated 1.3 million motorcycles. Ho Chi Minh City in Viet Nam and Penang in Malaysia have about 300 motorcycles per 1000 persons. The number of these vehicles is expected to grow very rapidly in China, Viet Nam, India and other low-income countries; for example, it is projected that there will be 70 million motorcycles in China by 2015.

Unmanaged growth of motorization is the root cause of many of today's urban transport problems. Due to imperfect systems of transport pricing, prices do not reflect the true cost of the provision of the transport services and facilities. Consequently, this has led to a waste of resources, insufficient funds to develop and maintain infrastructure, distortions in modal choice and the generation of externalities (pollution and congestion). Ideally, an efficient pricing system should be in place to realise the full cost of travel from the motorists to rectify the current situation. Alternatively, serious consideration needs to be given to the introduction of measures that include, inter alia, restraint and demand management strategies to control the growth and usage of motor vehicles, particularly the usage of private cars. The rapid rise in motorization has led to major problems with congestion in most growing cities of the region.

The central parts of many capitals, such as Bangkok, Delhi, Dhaka, Jakarta, Metro Manila, and Seoul, are particularly congested, with weekday peak-hour traffic speeds reported to average 10 km per hour or less. One estimate put the average travel time for work trips in Asia at 42 minutes. In large cities this number can be much higher, as in the case of Bangkok, where the average is estimated to be about 60 minutes. Delays due to congestion account for a significant proportion of the total trip time.

The estimated social cost of congestion could be enormous. A study in 1995 estimated the direct economic costs of congestion in Bangkok at 163 billion baht annually. The total cost represented 27 billion for the additional costs of vehicle operation, 20 billion for additional labour costs, and 116 billion for passengers' lost time. The total cost, however, did not include the cost of damage to the environment or human health. In many cities the level of congestion is so high that even a moderate reduction could provide significant benefits. A recent World Bank study estimated that a 10 per cent reduction in peak-hour trips in Bangkok would provide benefits of about US$ 400 million annually.

Most countries have addressed the challenge of increasing motorization through a combination of increased investment in urban road stock and the development of complementary public transport initiatives.

Public Transport Initiatives

Public transport has a very important role in urban transportation. Many cities such as Hong Kong, China; Singapore; and Tokyo, where the modal share of public transport is 70 per cent or more of total person trips, are deemed to be public-transportation oriented. In

Bangkok, Jakarta, and Manila, the modal share of public transport varies between 40 and 60 per cent of total person trips. In most cities of the region, the majority of the common people, the poor and other disadvantaged groups are very heavily dependent on public transportation.

Compared with private cars, public transportation is more sustainable on economic, financial, social and environmental grounds. However, the failings of public transportation have become one of the major challenges faced by many cities. Dissatisfaction with the level and quality of public transportation services leads those people who can afford it to turn to private modes of transport. Another common problem in many cities is that women, people with disabilities and other disadvantaged groups have poor access to public transport services and that it is found difficult to meet their basic mobility needs.

Urban Rail Projects

Governments in many countries have begun studying or implementing projects to develop rail-based transit systems in response to the shortcomings of road-based transport systems to meet growing demand in very large cities. Several Asian cities have addressed this issue through the implementation of mass transit systems. Underground rail systems are a long-established feature of Asia metropolises such as Tokyo; Hong Kong, China; and Seoul. These systems continue to be enhanced and improved.

For example, Hong Kong, China's Mass Rail Transport system was recently enhanced by the Tseung Kwan O Extension project. This project, which commenced operation in 2002, is the Mass Transit Railway Corporation's sixth rail line. Prioritized under the government Railway Development Strategy in 1994, the 12.5-km extension serves the rapidly developing Tseung Kwan O and Yau Tong areas.

The new project includes three main features: a diversion of the existing Kwun Tong Line from Lam Tin Station, passing through Yau Tong and terminating at Tiu Keng Leng Station; the new Tseung Kwan O Line connecting with the Eastern Harbour Crossing and passing through Yau Tong towards Tseung Kwan O New Town; and a fork in the line from Tseung Kwan O leading to termini at Po Lam and Tseung Kwan O South. The initial phase will include five stations Yau Tong, Tiu Keng Leng, Tseung Kwan O, Hang Hau and Po Lam as well as a depot. The Tseung Kwan O Extension project comprised 33 major contracts at a total value of $ 21 billion. Of these, 13 contracts were for civil works and the remainder are for electrical and mechanical works.

Two projects in China are enhancing this country's railway system. In preparation for the 2008 Olympics, Beijing has a major new investment plan for new metro and light rail transit lines of about 160 km. Shanghai is extending its current lines and building new lines including an elevated light rail train on the Pudong side. In Tehran, which has a network of 61 km (3 lines) with 32 stations, Lines 1 and 2 are being extended. In Mashad in the Islamic Republic of Iran, a citywide light rail transit system is under construction. It is 18 km long and will have one Line 1 with 22 stations. Manila is extending its present network. A major extension of Line 1 is in the approval process.

The Express Rail Link (ERL) in Kuala Lumpur opened in April 2002. An electrified linkage which is built on standard gauge to enable it to operate at 160 km per hour, the express railway links Kuala Lumpur International Airport with Sentral Station with two types of passenger service. One of these is the Airport Express service, which operates every 15 minutes and provides a non-stop journey time of 28 minutes. The other is a commuter rail service, which opened shortly after the airport express and serves three intermediate stations at Bandar Tasik Selatan, Putrajaya, and Salak Tinggi. This commuter rail service operates at 30-minute intervals providing an end-to-end journey time of 36 minutes. The station at Putrajaya that is currently under construction, is strategically positioned to support a new city by that name. The city of Putrajaya is part of a 15 by 50 km corridor designated for and designed to attract high tech multimedia industry development. Initially the rail link is expected to carry 6,000 passengers per day but as the new city grows the number is expected to increase to 10,000 passengers per day.

The new railway link has been financed, designed and built and will be managed, maintained and operated on a 30-year concession granted by the Malaysian Government to ERL SB, a private company which has two shareholders, Tabung Haji Technologies which has a 60 per cent stake and YTL Corporation which has the remaining 40 per cent stake. German banks and the German development bank KfW, and a combination of the Malaysian government and shareholders' equity shares have provided 50:50 funding for the project. A separate company, Express Rail Link Maintenance Support (EMAS) has been formed by Siemans, which has a 51 per cent share and ERL SB with a 49 per cent share to operate and maintain the link under a Ringgits 103.5 million contract.

Sydney Metro Light Rail is owned by the Sydney Light Rail Company, whose shareholders include the Australian Infrastructure Fund, the Utilities Trust of Australia and the Colonial Group. The first stage of the Sydney Metro Light Rail line was established in 1997 as part of the urban renewal of the Ultimo/Pyrmont area. Stretching a distance of 3.6 km from Sydney's Central Station to Wentworth Park, the line provides 10 stopping points through key tourist sites including Chinatown and Paddy's Markets, the Entertainment Centre, Sydney Convention and Exhibition Centre, the National Maritime Museum, Sydney Casino and the historic Pyrmont Pennisula. In August 2000 the first extension of the service into the inner Western Sydney suburbs of Glebe and Lilyfield included four more stops and increased the distance of the line to nearly 7 km. The inner western extension operates with the original set of seven Varioatram trams with services provided at 10-minute intervals.

The construction costs of $ 20 million for the western extension were shared between Sydney Light Rail Company and the New South Wales Government, with the Government paying the lion's share of $ 16 million. Metro Light Rail (Connex) operates the light rail service on behalf of Sydney Light Rail Company on a seven-year management contract. Sydney Light Rail Company wants to extend the line a further 2.5 km into the inner west creating a link to a number of different amenities in the area and providing residents within the inner west a quick link from Central Station to the central business district to Circular Quay. But the New South Wales Government has put the plans to do so on hold.

The first rail transit system in Bangkok has been in operation since 1999. Known as the "Skytrain", it is a 23-km elevated heavy rail transit system. The system has 23 stations and serves the inner areas of Bangkok. It has a maximum capacity of 25,000 passengers per hour per direction, with a 3-car train configuration at present. The capacity can be doubled in the future with 6-car trains. A private sector operator has built the US$ 1.22 billion mass transit system as a BOT project under a 30-year concession. An extension of the system with some support from the public sector is now under construction.

An underground rail transit system of similar capacity is under construction. This 20 km long underground system, with 20 stations, is scheduled to go into operation in April 2004. It is being built by a state agency, but will be operated by a private operator under a 25-year concession agreement. When completed, the two rail mass transit systems together will form a loop around the central area of the city. Bangkok also has a plan for light rail systems to serve the city's suburban areas in the future.

In December 2002, the first five-mile leg of the Delhi metro system was opened. The system was used by over one million passengers on its first day of operation – well in advance of the design capacity of the system. It will be decades, however, before the US$ 100 million per km metro will serve more than a handful of all commuting trips. In the meantime, Delhi government officials are under pressure to show some immediate improvements for the rest of the public, which relies on buses. The idea of pushing high capacity buses, segregated bicycle facilities, and integration of vendors into the street system design, has been championed by the sustainable transport committee advising the Delhi Government which includes Dinesh Mohan of the Indian Institute of Technology. Finally, it seems, these voices are being heard.

In China, subway lines have also been built over the last decade in Beijing, Shanghai and Guangzhou costing an average of US$ 72 million to US$ 96 million per km. Shanghai also recently opened its US$ 1.6 billion magnetic levitation train, which runs from a new international airport into the city. Shenzhen has announced plans to invest RMB 69 billion (US$ 8.4 billion) in subway, urban rail and light rail projects between 2003 and 2010. The projects involve eight separate railway projects totalling 250 km in length, half of which are expected to be launched next year. Shenzhen, which has boomed in recent years, is suffering from severe congestion problems and desperately needs new public transport options. Other cities in which rail-based mass transit systems have either been implemented planned or under active consideration include Busan and Incheon (Republic of Korea), Kolkata (India), Manila (Philippines), Daegu and Tianjin (China), Bangalore, Dhaka, Hyderabad, Karachi, and Mumbai (India).

Improving Bus Networks

Although mass transit projects will play an increasing role in moving the populations of Asia's megacities, bus transport will continue to provide the most flexible and widely used means of public transport for the foreseeable future. For example, Tiwari reports that: "Buses form the backbone of the transport system in Delhi. Buses constitute less than one per cent of the vehicle fleet, but serve about half of all travel demand.... Bus service was expanded in 1996 by adding more buses, with buses per route increasing from 0.8 to 1.7. The regular fixed-route bus system now comprises about 4,000 privately operated buses and 3,760 publicly operated buses. 5,000 private charter buses that provide point-to-point service during peak hours to subscribers who pay a monthly fee for a guaranteed seat complement these initiatives. Schools and tourists use another 5000 buses. ..... Approximately 10,000 buses carry 6 million commuters along 600 routes everyday." Many city and national governments have developed plans for improving the efficiency of this vital component of the urban transport system.

Increased attention is also being paid to improving the comfort and amenity of bus services. Premium (air-conditioned) bus services are now available in a large number of cities in the region. Cities with relatively higher incomes such as Bangkok, Kuala Lumpur, Shanghai, and Shenzhen have introduced higher-quality buses on their roads. Advanced technology, low-floor kneeling buses have been introduced in Hong Kong, China; Singapore; and many Japanese cities to facilitate the embarking and disembarking of passengers, particularly for elderly passengers or passengers with disabilities.

Public Transport System Integration

Another major development is the integration of public transport services across modes. Cities with more advanced forms of transportation such as Singapore and Hong Kong, China have successfully integrated their various public transport services provided by multiple operators, such as the underground and bus systems. Seoul and Metro Manila, meanwhile, have been less successful in modal integration. Bangkok has prepared a plan to integrate the city's bus services with the rail transit systems, such as the underground system currently under construction.

Making the public transport system easier to use can encourage the use of public transport in preference to private cars. Singapore has been an international pioneer in this regard. Singapore has an efficient and reliable integrated public transport system comprising bus, metro and light rail services. Three private operators run the bus services. Together they operate 261 routes (175 trunk and 86 feeder) with a fleet of about 3,500 buses. The average daily number of passengers is about 3 million. The metro rail system was introduced in 1987 and has two main lines with 49 stations. It has an average daily ridership of about 1.1 million passengers.

The system is now being extended to connect to the airport and this extension has been in operation by 2002. Another feeder light rail system was introduced in 1999, mainly to complement the metro rail. A separate service company has been set up by the bus and metro operators to develop an integrated public transport system. Central planning and coordination of the bus network taking into consideration the metro and light rail systems has reduced wasteful duplication and improved transit services. The transit modes have a common ticketing system with a cashless mode of payment by stored-value magnetic cards. The ticketing system has been further improved by the introduction of a contactless smart card system.

Para-transit

Public transport in many Asian cities is characterized by a mix of formal public transport routes – often publicly operated – and a wide range of both motorized and nonmotorized conveyances available for hire to public. Less formal means of public transport play a particularly important role in low to middle income countries.

Dhaka has a low level of motorization. About 60 per cent of all trips are made on foot. Almost half of the remaining trips are made by rickshaw. Out of total vehicular persontrips, the share of public transportation is less than 25 per cent, of which the bus is the most popular mode. About 2,000 buses are operated, most of which are old and small. In the past few years, private sector operators have introduced about 200 air-conditioned buses.

Bangladesh Road Transport Corporation, a public sector agency, owns a limited number of single and double-decker buses. Private operators run these buses under a lease agreement with the agency and in competition with the private sector along the high-demand corridors.

The number of motorized and non-motorized three-wheelers has grown quickly as a substitute for buses, although they are expensive and costs are higher than for a feasible premium bus service. The non-motorized transport (NMT) modes, especially the rickshaws (the estimated number of which is more than 300,000), play a significant role. They effectively operate on almost all the roads of the city, except a few major roads. There are about 70,000 two-stroke engine three-wheelers and 3,000 taxis. These three-wheelers are the major sources air pollution in Dhaka. Due to poor traffic management, lack of enforcement of traffic rules, and operation of a diverse mix of modes sharing the same right, the city experiences chronic traffic congestion and other related problems on all major roads.

Bangkok has a large paratransit fleet of 49,000 licensed taxis; 7,400 3-wheeler tuktuks; 8,400 silor-leks (small 4-wheelers), and about 40,000 hired motorcycles (which provide services in lanes off the main roads). The majority of taxis and all tuk-tuks are LPG-powered. A recent innovation was the introduction of mini vans by the informal sector. About 8,000 14-seater minivans serve commuters on 103 routes, mainly between suburban locations and the central areas.

Non-motorized Transport

Non-Motorized Transport (NMT), including walking, remains a viable option to meet the basic mobility needs of all groups in a sustainable way. Unfortunately, NMT has received the least attention in traditional urban transport planning. Consequently, it has been either overlooked or totally neglected. Public policy towards NMT has not always been favourable and there is a need to revise public policies to promote the safe operation of NMT in Asian cities. Public policies and investments need to be directed to enable greater use of NMT by providing suitable rights-of-way (shared or exclusive) and pedestrian facilities, and by giving attention to safety issues.

In the Philippines the Global Environment Facility is implementing a component of the broader Metro Manila Urban Transport Integration Project (MMURTRIP), which is co-financed by the World Bank and the Government of the Philippines and aims to improve urban transport in Manila. This objective will be achieved through the improvement of street level interchanges between buses, jeepneys and Light Rail Transit (LRT), and the implementation of effective traffic management measures along major travel corridors.

GEF's component aims to promote a shift from motor vehicles to non-motorized transport (NMT), particularly bicycles, in Marikina City, metro Manila, by making NMT a safer and more convenient transport mode in the city. Through this shift to NMT, the project also aims to slow the growth of transport-related greenhouse gas emissions. To achieve these aims the project will focus on demonstrating the benefits and viability of NMT as an alternative transport mode, so as to encourage replication of this pilot NMT programme in other parts of Metro Manila, elsewhere in the Philippines, and in other countries.

The project includes the construction, evaluation and promotion of the Marikina Bikeway System (MBS)-a 66-km long network of trails and road lanes specifically designed for NMT, plus bicycle parking and traffic calming systems. The pilot bicycle network will connect low-income families and squatters in residential communities with schools, industrial employment centres, the new metropolitan train station and other public transport terminals. It is hoped that the new NMT-friendly facilities will encourage the use of NMT modes, and connection with the public transport terminals will promote the combined use of NMT and train/bus for trips between Marikina and the rest of metropolitan Manila.

The anticipated benefits of the project will be less motorized traffic and congestion and the consequent decrease in emissions of greenhouse gases and other pollutants. GEF estimates the reduction at more than 30,000 tons of carbon dioxide equivalent per annum. In addition an indirect benefit will be the demonstration of the advantages of bicycle and other non-motorized transport as a model for other places in Metro Manila and the Philippines.

Case Studies: Transportation in Surabaya, Delhi and Hanoi

Case studies on urban transport development in Surabaya, Colombo and Hanoi are presented in this section. These three cities are not representative of the whole range of urban transportation situations in the ESCAP region; their selection was rather determined by data and information availability.

Surabaya

Surabaya is a city with a 1990 population of 2, 473,722 people, comprising 374 sq km in area, on the east coast of the island of Java. It is 17 km north to south, and 22 km east to west. The city has a one-way traffic system, allowing for relatively high traffic speeds. Direct route trips have been sacrificed for the system. "The impact of the one-way system on motorized vehicles alone is to increase daily passenger car unit km travelled by 7,015 km, and increase travel times by 265 hours per peak hour on an average day".

There has been little consideration for public transportation or for non-motorized road users. Vulnerable road users constitute 27 per cent of reported accident victims, and motorcyclists' account for another 46 per cent of victims. Vehicle ownership has increased dramatically. By international standards, Surabaya has an extremely high mode share of private motorized trips, predominantly motorcycle, relative to per capita incomes, despite the fact that average trip distances are extremely short.

To overcome the serious issues facing Surabaya the GTZ's Sustainable Urban Transportation Project in Surabaya was initiated. The project aims to work with related agencies and the people of

Surabaya to devise and implement policies to establish a sustainable urban transportations system. The project is expected to provide a model of how to reduce carbon dioxide emissions from the transport sector in large cities in developing countries.

There are five key areas of focus: access; equity; pollution prevention; health and safety; and public participation and transparency. Essentially the strategy aims to develop Surabaya into a city that can meet the needs of a broader range of road users other than just vehicles. It aims to ensure that the people of Surabaya, both the financially resourced and those will little or no access to financial means, are able to participate in the economy of the city.

A range of short, mid and long-term goals have been identified to bring about the realisation of the strategy. For example in the short term the goals are:

- Road accidents causing fatalities and injuries must be reduced by 30 per cent;
- The first major improvement in public transport must commence, through improving bus service reliability, and the introduction of a 'green route' (a route only operated with low emission and low floor buses); and
- An integrated bicycle and becak (non motorized pedicabs) lane must be built. Construction of more roads is not part of the sustainable urban transport system for Surabaya. It recognized that more roads lead to more usage, more congestion, greater pollution and a tendency to meet the needs of the affluent that have the capacity to own vehicles.

The main concerns of sustainable urban transport are public transport, non-motorized transport, modern and clean technology, application of economic and fiscal instruments, and institutional reforms and public participation.

Delhi

In Delhi, alarming concern over the high level of respirable particulate matter has triggered strident public campaigns catalysing judicial action in Delhi. Spurred by the Supreme Court directives Delhi has taken more advanced action than any other city in India to cut vehicular emissions. The actions include implementation of the Euro II emissions standards, lowering of sulphur to 500 ppm in fuels, lowering of benzene to one per cent in gasoline, moving its entire bus fleet, three-wheelers and a large numbers of taxis and mini buses to

CNG, capping the age of commercial vehicles at 15 years, freezing the numbers of three wheelers, and finding ways to take transit traffic away from the city. Successful implementation of the CNG programme especially the replacement of all diesel buses with dedicated CNG fleet to cut diesel particulate emissions is very significant. The World Bank study of 1998 had estimated that diesel vehicles contributed nearly 65 per cent of the total particulate emissions from mobile sources in Delhi. Currently, there are as many as 80,103 CNG vehicles in Delhi-the largest fleet ever in the country. Around 9,044 CNG buses in all are the principal mode of public transport in the city today.

New Delhi has largely focussed on cutting emissions at source by pushing for improvements in technology and fuel quality. As a result, the city has just about succeeded in stabilising pollution levels. The recent release from the Central Pollution Control Board reports 24 per cent drop in PM10 levels between 1996 and 2002.130

However, rapid growth in the use of cars and motor cycles, setting in motion a neverending spiral in demand for more road space points to an even bigger challenge in controlling the pollution levels. The road network has increased nearly three times-from 8,380 km in 1971-72 to 28,508 km in 2000-01. But during the same period vehicles have increased a staggering 16 times, and in particular the number of cars in the city has been increasing at a much faster rate than the number of two-wheelers. A survey conducted by the Government of Delhi in 2002 showed that while cars make up about 26.6 per cent of vehicles in Delhi they only cater for about 7 per cent of the city's travel demand. By contrast, buses account for no more than 1.2 per cent of the total number of vehicles but cater for 60 per cent of Delhi's travel demand An earlier study reinforces the fact that the use of bicycles has fallen significantly as a source of transport for Delhi's population, falling from 36 per cent of vehicles in 1957 to a mere 6.61 percent in 1994.132

By the mid-1990's Delhi had 1,749 km of road length per 100 sq km area by comparison with the national average of 73 km per 100 sq km area. Nevertheless, the amount of road space is not enough to satisfy the traffic demand and so the average speed of vehicles, which hovers between 21 to 39 km per hour and sometimes drops to about 6 km per hour, is steadily going down causing more congestion, pollution and fuel loss.

Delhi Government announced a transport plan for the city in September 2002. The action points most relevant to public transport include rationalisation of bus routes, bus lanes for selected corridors, introduction of premium bus services, time table integration of bus and metro as the short term and providing parking facilities. The medium and long-term plans include running a high capacity bus system (on five selected corridors to begin with), electric trolley buses (on two selected corridors to start with), and feeder bus route for the metro and finalisation of proposals for bus lanes and bus only routes.

This plan has yet to take off. There is still no clear projection of how the various forms of public transport would be augmented to meet the overall transport demand and the likely impact of these on the usage of personal transport and overall emissions. Currently, there is only an 8 km stretch of rail-based metro that is operational and an aggregated 198.5 km of metro is targeted for completion during 2015-2021. Environmental benefits of metro in terms of reduced emissions and congestion are not yet clear. According to the feasibility study on the metro rail in Delhi by Rail India Technical and Economic Services (RITES) the modal share of cars and two wheelers is likely to reduce only marginally whereas that of three wheelers and taxis will remain unchanged as these are on road by choice.

RITES claim that the emissions reduction potential of metro can be as high as 50 per cent. But clearly, that is possible only if fiscal and command and control measures are put in place to discourage personal transport.

Hanoi

As a result of the acquisition of new and larger buses by the city owned company TRANSERCO, as well as the existence of an attractive fare system and the regular availability of services, the number of passengers using public transport in Hanoi increased between 500 per cent and 2,000 per cent between 2001 and 2003. This increased usage of public transport has placed a number of pressures on the city's bus system including overcrowding of buses and bottlenecks at stations. The city is responding to these challenges by:

- Encouraging private companies to build and operate up to 40 per cent of the public transport services thus releasing capital funds for expansion of infrastructure and services;
- Creating a system of corridors with dedicated bus lanes and accessible bus stops and interlinking the corridors so that passengers can traverse the city through a number of destinations safely and efficiently;

- Developing some of the corridors in such a way that they can be upgraded to accommodate urban rail lines and streetcars;
- Developing the Hanoi urban railway system and integrating it into the public transport system, with tickets valid on either bus or train;
- Implementing a new ticket control system by early 2004;
- Improving the comfort and environmental characteristics of the city's bus fleet; and
- Extending operation times and schedule intervals and improve public information about the system.

The Traffic Management Centre as the regulating authority will develop the regulatory framework within which private companies will be invited to operate dedicated bus lines through a management contract. The Public Transit Authority will have responsibility for allocating the lines as well as fixing fares. The city intends to establish no more than two management contracts in 2003 with a maximum of five companies ultimately participating in the system. To fully implement the planned bus system some corridors will only require minimal improvements to intersections and available road space, while others will require significant road and bridge construction and resettlement of people occupying land needed for the corridors.

Currently about 80 million passengers per annum use Hanoi's public transport system but by 2020, it is estimated that about 250 to 300 million passengers will use the system. The goal of the plan is to be able to accommodate this increasing demand through improvements to bus ways and their integration with urban rail.

By that time the city hopes to have three to four companies providing these services on the basis of formal environmental and safety standards, and improved driving and mechanical maintenance skill levels and licenses for bus drivers.

7

Transport and Environmental Degradation

Transport related activities have many impacts on the environment. The most important negative effects are contribution to climate change and to local air pollution. The transport sector is also responsible for increased noise levels, acidification, eutrophication, habitat loss, water pollution, and waste generation. Natural resource depletion and negative visual effects are other consequences of transport related activities. In addition, transport infrastructure may significantly affect social and economic factors in local communities and influence people's health and safety.

Road transport is responsible for the majority of negative impacts of the transportsector on the environment. In OECD countries, road transport accounts for 80% of theenergy consumption in the transport sector (OECD, 2001). Air transport also accountsfor a great share of negative environmental impacts, while rail transport causes muchless damage (OECD, 2001). In 1997 in OECD countries, motor vehicles accountedfor 89% of CO, 52% of NOx, and 44% of VOC emissions (OECD, 2001).Environmental impacts of transport can be direct, indirect, and cumulative (Tsunokawa & Hoban, 1997). Direct impacts are, usually visible changes in the environment caused by constructing the transport infrastructure. For example, land consumption and removal of vegetation belong to this group. Direct impacts are easier to prevent and deal with, since they are usually obvious.

Indirect impacts, on the other hand, are less obvious, more difficult to predict and to measure, and hence usually cause more serious damages to the environment.

For example, erosion of adjacent soil caused by road construction can, over time, lead to a serious pollution of watercourses. Another very common indirect impact caused by expansion of the transport infrastructure is an increasing stress (through, for example, urban development, logging, and hunting), on adjacent ecosystems now easily accessible to people. Cumulative impacts can be caused in different ways: by accumulating impacts of several smaller interrelated projects, by a sudden catastrophe, or by slow incremental negative changes (Tsunokawa & Hoban, 1997). Cumulative impacts are results of additive or synergistic effects that lead to serious damages in one or more ecosystems.

Environmental Impacts of Transport

Natural Resource Depletion

Fossil fuels are the primary energy source for transport. Only in OECD countries, the use of fossil fuels for transport increased by more than 45% from 1980 to 1997 and is expected to continue growing (OECD, 2001). To be constructed, transport infrastructure requires a substantial amount of concrete and steel. In order to produce vehicles, metals and plastic are required. The extraction and production of all these materials cause damage to the environment.

Impacts on Air Quality

The transport sector, especially road and air transport, contributes to air pollution, acidification and climate change through emissions of carbon monoxide (CO), carbon dioxide (CO2), nitrogen oxides (NOx), hydrocarbons (HC) particulate matter (PM), lead (Pb), heavy metals, and volatile organic compounds (VOC).

These pollutants are released during the combustion of fossil fuels, the primary energy source for transport. The most important air pollutants will be presented in the following sections.

Primary Air Pollutants

Nitrogen Oxides (NOx)

Nitric oxide (NO) is the major gas released during fuel combustion under high temperatures and pressure. This gas, released in the atmosphere, is usually quickly oxidized in NO2. NOx, together with SO2, play an essential role in acidification. NOx also react with hydrocarbons, producing photochemical smog. These reactions are stimulated by the sunlight.

Hydrocarbons (HC)

Hydrocarbons include several hundreds organic substances, created during the incomplete combustion of fossil fuels. The most important hydrocarbons are benzene and ethylene. As already mentioned, together with NOx, HC create photochemical smog.

Carbon Monoxide (CO)

Carbon monoxide is also created during the incomplete combustion of fossil fuels. The main source of both CO and HC are gasoline engines. Diesel engines produce much smaller quantities of these pollutants. CO is highly toxic for people, because it bounds to the blood hemoglobin, lowering its capacity to carry oxygen. It also has negative impacts on heart, circulation, and nervous system.

Carbon Dioxide (CO_2)

Carbon dioxide is released during the combustion of fossil fuels. Its emissions are directly dependent on the quantity of the fuel burned, because there is no available technology for its subsequent removal. The only way to lower CO2 emissions is to use fuels with less carbon content or to lower fuel use by improving energy efficiency (CEI, 1999).

Sulfur Dioxide (SO_2)

Sulfur dioxide emissions mostly originate from diesel motors. The emission of SO2 depends on the sulfur content of the fuel. SO2 contributes to acidification, together with NOx.

Particulate Matters (PM)

The particulate matters include many different substances that have different origin. The common characteristic is that they all contain a carbon nucleus and various other components (hydrocarbons, inorganic sulfates and nitrates, as well as metals and polycyclic aromatic hydrocarbons) adsorbed on it. The most common PMs are those that are created during the combustion of diesel fuel, materials that originate from tires, brakes, as well as dust.

Lead (Pb)

Lead is emitted into the atmosphere in form of fine particles. It is a typical constituent of leaded gasoline, added in order to raise the octane number and lubricate the engine. Lead is a neurotoxin, which has negative impacts on neurological development of children, and also causes cardiovascular problems for adults.

Aldehydes

The aldehydes are mainly produced by the combustion of alcohols and diesel fuel. Gasoline combustion emits small amounts of these substances.

Secondary Air Pollutants

The secondary air pollutants are substances created in chemical reactions among primary pollutants. For example, NOx and HC create ozone (O3) in the presence of sunlight.

Air pollution is a very complicated process that depends on many factors. The emission is determined by fuel composition (sulfur and lead content), engine maintenance (filters, pollution control devices, fuel systems), vehicle age (older vehicles have higher emissions), engine temperature (catalytic converters do not work before the engine reaches normal operating temperature), road geometry (decreasing and increasing the speed causes higher emissions), type of vehicle (large engines pollute more; gasoline engines emit more CO and HC; diesel engines emit more PM, SOx, and NOx), and speed and congestion (most vehicles are most efficient at speeds between 80 and 100 kilometers per hour) (Tsunokawa & Hoban, 1997).

Air pollution is rarely localized. How far will the pollutants be dispersed depends on several factors. The most important are: prevailing wind direction (concentration of pollutants is higher downwind of the road), weather conditions (rainfall, humidity, temperature), vegetation (filters pollutants) and topography (as physical barriers to pollutants) (Tsunokawa & Hoban, 1997). The most serious problems arising from emissions of pollutants in the air are climate change, acidification, and urban air pollution.

Effects of Air Pollution

Climate Change

The climate change problem is related to changes in the concentration of the greenhouse gases (water vapor, CO2, CH4, N2O, and CFCs), which trap infrared radiation from the Earth's surface and thus cause the greenhouse effect. This effect is a natural phenomenon, which helps maintain a stable temperature and climate on Earth. Human activities, such as fossil fuel combustion, deforestation, and some industrial processes have led to an increase in greenhouse gases concentration. Consequently, more infrared radiation has been captured in the atmosphere, which causes changes in the air temperature, precipitation patterns, sea-level rise, and melting of glaciers. Carbon dioxide (CO_2) is the most important greenhouse gas produced during the combustion of fossil fuels for transport activities.

Acidification

The problem of acidification is caused by acid depositions which originate from antropogenic emissions of the three main pollutants: sulfur dioxide (SO_2), nitrogen oxides (NOx), and ammonia (NH3). Acid depositions have a negative impact on water, forests, and soil. They cause defoliation and weakening of trees. Changes in soil and water pH have a harmful effect on soil and aquatic organisms. Damage is also visible on man-made structures, such as limestone and marble buildings and monuments. The main sources of emissions of acidifying substances are coal and other fossil fuel combustion used for energy production and transport, as well as use of animal manure in agriculture.

Urban Air Pollution

Road transport is the main contributor to air pollution at the local level in urban areas. Negative effects on human health are mostly caused by nitrogen oxides and particulate matters emitted from vehicles. These gases, along with non-methane volatile organic compounds (NMVOCs), methane (CH4), and carbon monoxide (CO) in complicated photochemical reactions generate photochemical smog. Tropospheric ozone is the main component of photochemical smog and the major air pollutant in urban areas in temperate regions. It causes respiratory and eye irritations, headaches, and can cause respiratory diseases, including asthma. Ozone also has negative impacts on vegetation. Despite all the improvements in the motor vehicle technology in industrialized countries, smog is still a problem in large cities. In developing countries, the situation is even worse because of old vehicles and poor maintenance (Pastowski, 2001).

Land Use

It is estimated that transport infrastructure consumes 25-30% of land in urban areas in OECD countries (OECD, 2001). In the EU, 93% of total land area used for transport belongs to roads, while rail and airports occupy 4% and 1% respectively (OECD, 2001). Increased land use for transport infrastructure increases pressures on biodiversity due to habitat fragmentation. It also causes an increase in acidification and eutrophication. Land is affected by the transport sector in two ways: directly through building the transport infrastructure, and indirectly by the development induced by the transport sector (EC, 1999).

One of the most obvious negative effects of the transport infrastructure development on land use is urban sprawl. The growth of urban areas over the surrounding rural land leads to fragmentation of land use control among more localities and segregation of types of land use in different zones. Urban sprawl causes other problems, such as widespread strip commercial development, low-density settlements, and dominance of private motor vehicles in the transportation modes (Pastowski, 2001).

Impacts on Soils

An unfortunate fact is that the soil best for building the transport infrastructure is alsobest for agriculture, because it is stable and flat. Therefore, transport infrastructure development inevitably leads to the loss of reproductive soil for agriculture, and thus causes damages to the socioeconomic development of an area. Not only does the soil covered by the transport infrastructure become lost, but also adjacent soil, which is damaged by the construction works as a result of compaction by heavy machinery. Transport infrastructure construction often requires at least a partial clearance of vegetation. This often leads to erosion as an indirect effect of construction. In some cases, erosion may occur far from the transport infrastructure that actually causes it, as a result of cumulative impacts.

Pollution of soils in close vicinity of roads by chromium, lead, and zinc, may be are sult of a very busy traffic. These metals tend to remain in the soil for several hundred years and cause damage to the soil microorganisms and vegetation. Fortunately, these effects are localized on the narrow area on both sides of the road.

Impacts on Biodiversity

There are three ways in which the transport sector contributes to biodiversity loss: direct damage, fragmentation, and disturbance (EC, 1999). Loss of habitat is an inevitable consequence of land use change during the construction of the transport infrastructure. However, by careful planning, it is possible to keep the damage at an acceptable level. If the construction is not carefully planned, especially in sensitive areas, it can destroy or seriously damage natural ecosystems, thus causing direct damage through loss of habitats for sensitive plant and animals, which is the main cause of biodiversity loss.

Roads cause fragmentation of habitats, preventing free movement of animals and exchange of genetic material. Habitat fragmentation damages ecosystems' stability and health. Habitat fragmentation can cause corridor restrictions. Corridors are routes that animals use for satisfying their everyday or seasonal needs for food, breeding, and shelter. By cutting through the corridors, the transport infrastructure causes negative pressures on animal populations affecting their feeding or breading, because they are either reluctant to cross the roads or get killed while crossing it. It is also a case that some animals are attracted to roads for various reasons-more food, shelter from predators, or easier movement-which often leads to increased mortality due to accidental deaths. Road construction also opens the ways for intruding species, disrupting in this way the ecological balance of the ecosystems. Noise, lights, and runoff of hazardous compounds from roads cause disturbance in the ecosystems, and lower there production rates of animals (EC, 1999).

Water ecosystems also suffer disruptions caused by the land transport infrastructure. Erosion leads to accumulation of fine earth particles downstream, which affects habitats for fish spawning. Changes in water flow caused by diversions during construction works often have negative effects on plankton, upsetting eventually food chains in the ecosystem. Roads can also cut through the migration routes of fish, causing disruptions in the spawning cycle.

Impacts on Water

Activities caused by the transport sector cause surface and groundwater flow modifications, as well as water quality degradation. Modifications in the flow of surface waters are caused by diversions of water flows, which contribute to flooding and soil erosion that often happen far from the place of diversions and the road itself.

Groundwater is often affected by road constructions, such as drainage and embankments. Changes in water tables negatively affect vegetation, increase risk of erosion, and often cause loss of water for drinking and agriculture. Modification of the flow of surface and groundwater has a negative effect on fish and other animals. Transport causes pollution of water bodies adjacent to transport infrastructure. Runoff from roads contains hydrocarbons, heavy metals, chemicals used for de-icing, and other chemicals. Railway power lines release copper. Effluents from ships cause water pollution. Transport of dangerous goods (hazardous wastes, oil) poses a risk of contamination of soil, waters, and wet lands.

Ground water quality may also be affected by leaks from the underground fuel storage tanks (UPS) constructed in connection with filling stations. Transport by water affects coastal zones through building the port infrastructure. High-speed ships can cause a serious disturbance in sensitive areas of rivers and seas. The transport of oil and chemicals poses a risk of accidental water and coastal pollution.

Noise

Noise is probably the most obvious impact coming from the transport sector. Excessive noise levels (65dB(A)2 and higher) damage people health by contributing to high blood pressure and cardiovascular diseases (OECD, 2001). In OECD countries, around 30% of the population is exposed to noise levels higher than 55dB(A) (OECD, 2001). Road noise comes from four sources: vehicles (engine work, acceleration, braking); friction between vehicles and road; driver behaviour (horn usage, loud music, shouting, sudden braking or start); and construction and maintenance work (heavy machinery) (Tsunokawa & Hoban, 1997).

Continuous noise, even if its levels are not too high, increases stress levels by causing annoyance and disrupting communication among people. Continuous exposure to noise can lead to weakening of the auditory system and sleeping disorders. Noise has negative affects on wildlife; animals are often afraid of noise and do not approach roads, which can disturb their breeding, feeding, or migration patterns. Another negative effect related to transport is vibration. Vibration, mostly caused by road freight transport and air transport, is very damaging to lightly built structures along the road, as well as cultural heritage monuments. Vibration can also have negative impacts on people, causing sleeping problems and general disturbance of normal living patterns.

Visual and Aesthetic Impacts

Visual impacts represent the blocking out of light and pleasant views by the transport infrastructure and activities, while aesthetic impacts are concerned with the actual design and style of the transport infrastructure (Button & Rothengatter, 1993). Negative visual and aesthetic impacts of the transport sector are the consequences of poor planning, without consideration of the main landscape design principles. Some of the principles will be explained below.

The road must be in harmony with the landscape. This means that it should not take control over the landscape, but try to coexist with it. The road must follow the relief and morphology of the landscape as closely as possible. It is necessary for the road to be well visually incorporated into the landscape; it must not block or cut off a view that is of great aesthetic, natural, historic, cultural, or archeological value.

Constructing the transport infrastructure requires careful consideration of watercourses and vegetation. Significant amount of water diversion or deforestation should be avoided by finding alternative routes that show more respect for nature. Urban planning and transport infrastructure construction must be considered together. It is very often the case that road construction induces urban development. Sometimes, however, this development may be undesirable and may have negative visual and aesthetic impacts.

Social and Other Negative Impacts of Transport

Impacts on Communities and Economic Activity

Although the transport infrastructure intends to connect people and increase the speed of their communication, in some cases, if not planned carefully, it can cause the opposite. Building a highway, for example, over the existing routes between settlements and commercial areas can change the routes the communities used before to reach shops or schools, because people may be reluctant to use the highway crossings if it requires more time and effort than they are ready to spend.

Another example is a road built over one of two fields a farmer works on. First, the farmer loses a part of his land because the road covers it. Second, the road crossing is situated far from both fields, so the farmer needs to invest a considerable time and effort to go from one field to the other. In both examples damage is being made to both people, who will suffer the loss of job or income and will need to change their habits, and the economy, since the changes in travel routes and community interactions will inevitably lead to losses in the economic sectors.

Even widening of roads can have negative impacts on the communities and economy. Usually, roadsides are places of very active social and business life (shops, restaurants, and cafes).

The widening of roads inevitably leads to loss of business and customers for the owners and disruptions of living habits for their customers. A road bypassing the community is sometimes a good solution. It preserves local modes of communication and does not cause losses for the economy. However, it may happen that, in order to attract more customers, some businesses migrate from the community to the areas closer to the road. In that case, the community will suffer losses. It is, therefore, very important to carefully weight both options and decide whether it is really better for the community to be bypassed by the road.

Impacts on Human Health and Safety

Roads contribute to air pollution in the local areas. They are also corridors for transmission of diseases between local population and construction workers, and also between plants and animals. Road construction period is time of high risk for transmission of diseases. Construction workers often get endemic diseases, which they later transfer to other areas. Road construction sites represent great opportunity for development of water borne diseases, due to poor sanitation conditions. These sites are also a potentially good environment for transmission of sexual diseases. Roads represent a source of noise and vibrations, both during the construction and use.

Transport infrastructure is associated with a rather high risk of accidents and injuries. Road accidents are a major problem, causing losses of lives and significant costs to the economies. Accident rates are decreasing in developed countries, but increasing in the developing world (Tsunokawa & Hoban, 1997). The most vulnerable groups of road users are pedestrians and users of non-motorized vehicles.

Environmental and Social Externalities from Transport Activities

A study conducted by WHO has shown that road transport is the major source of human exposure to air pollution and noise (OECD, 2001). It is estimated that total environmental and health costs from transport (air pollution, noise, climate change, and accidents) represent almost 8% of GDP in European countries, road transport being responsible for 92% of the costs (OECD, 2001).

The costs of the environmental effects of the transport sector are difficult to calculate, since they are non-marketable and therefore represent externalities, "...effects where the profit or usefulness of somebody is affected by the actions of somebody else without any payment being received by the person who suffers the damage from the person who causes it." (Quinet, 1993).

Another difficulty in calculating the costs of the damage to the environment is a very large span of time over which the environmental effects occur. The transport infrastructure stays in the environment sometimes for centuries, causing negative impacts as long as it exists. The consequences of global warming, as another example, will last for many generations.

Finally, many environmental effects caused by the transport infrastructure are difficult to calculate, because it is impossible to estimate their future behaviour with certainty. For example, it is not possible to know with certainty what will the effects of global warming be in one hundred years from now, as well as what technological improvements will happen in the transport sector that will have an impact on the emission of the greenhouse gases.

Role of Transport in Development

It is widely acknowledged that transport has a crucial role to play in economic development. More specifically, it has been recognised that the provision of a high quality transport system is a necessary precondition for the full participation of remote communities in the benefits of national development.

The direct impact of transport on production at remote locations is derived from three effects:

- Lowering of production costs;
- Increased producer prices; and
- Encouragement of investment.

Lowering of Production Costs

The reduction in costs results from three main factors. Firstly and most obviously, improved transport lowers the delivered costs of inputs to the producer. This can be important for agricultural as well as industrial production: Ahmed and Hossain, in a study of two groups of villages in Bangladesh, found that agricultural output was 31 to 42 per cent higher in the group with better transport access, and attributed this difference principally to the lower delivered cost of fertiliser.

A second and related issue is the reliability of transport services. The importance of continuity of input supply increases rapidly as the degree of industrial sophistication increases. The absence of regular and reliable transport services operating with adequate frequency will effectively condemn remote communities to subsistence production in perpetuity. As shipping services generally use a larger unit of supply and operate at lower frequencies than land transport services serving markets of a similar scale, interruption to supply is generally a far more serious problem where the remote community is dependent on maritime transport.

Finally, improved transport can broaden the labour pool to which a production facility has access. While access to unskilled labour may not be a problem in most remote island communities, access to skilled labour frequently is.

This applies to both labour that is required on a temporary basis – for example, to the services of specialist advisers – and to skilled workers required for permanent employment. In Indonesia, the latter is likely to become increasingly important with the change in strategy in transmigration efforts which have recently 'focused not on moving people but on making locations more attractive and viable so that people want to move there themselves'. An important component of making remote island locations more attractive to potential migrants – particularly skilled workers – will be a reduction in the sense of physical isolation associated with them.

Increased Producer Prices

For many agricultural commodities and low value added manufactures, the costs of transport represent a substantial proportion of total product costs. One study has indicated that, in developing countries, transport costs typically account for between 10% and 30% of final product price.

Frequency and reliability of transport also have a very significant impact. Irregular or infrequent transport services require purchasers to hold high levels of stock in order to ensure that they in turn can ensure continuous supply to their customers. This results in an increase in inventory costs, which in turn depresses the prices offered to producers in remote locations. Added to this is the risk of spoilage of perishable products. This may seriously inhibit the diversification of primary activity into higher value lines such as horticultural production. Alternatively, it will significantly erode the benefits to producers of diversification into higher value but more perishable commodities.

Increased Investment

The quality of infrastructure and support services has been identified as a significant determinant in investment decisions. Creightley reports that 'for countries in the early phases of development, good quality infrastructure was preferable to tax incentives for attracting foreign investments'. Creightley also reports evidence that 'transport improves access to institutional credit, contributes in shifting the allocation of credit from nonproductive to productive activities, and leads to increased demands for credit'.

Virtuous Circle Effects

Transport sector improvements can serve as a catalyst that promotes a virtuous circle of economic development. The reduction in input costs and improved producer prices lead to improved profitability of agricultural and industrial production, creating an incentive to increase output. At the same time, greater access to investment funds permits the expansion of capacity required to enable producers to expand production in accordance with this incentive, and also facilitates upgrading of the technology of production. Economies of scale combine with improved productivity from capital deepening to further improve margins, and provide additional impetus for investments.

Increases in levels of production bring with them increased demand for transport services, improving profitability and encouraging further investment in transport itself. This in turn leads to improved service frequency and larger scale units of production (ships in the case of maritime transport), providing a basis for the next cycle of improvements in the agricultural and manufacturing production of the regions served.

The Palm Oil Industry

According to a World Bank survey, Indonesia is one of the world's lowest-cost vegetable oil producers (after soybean oil from Argentina and Brazil). Direct costs of production are far lower than international palm oil prices, and land costs are low. However, investment – and especially foreign investment-in the oleochemical industry in Malaysia, where production costs are much higher, has far outstripped investment in Indonesia. A recent industry study suggests that this is because 'some top managers from foreign companies which have palm oil business in both Malaysia and Indonesia evaluate that Indonesia's advantages in labour and land are offset by overhead burdens so there is no difference in total costs between the two countries.' The study authors identify the major disadvantages perceived by the\ industry side as possible bottlenecks inhibiting future development. The first two of these are:

- A shortage of port and storage facilities for palm oil products; and
- A shortage and poor maintenance of inland and offshore transport systems.

Transport and Personal Welfare

The contribution of transport performance to regional economic development has obvious implications for poverty alleviation and personal welfare. In addition, however, transport system performance can have a direct and significant impact on a range of other dimensions of development.

Health

In Indonesia, considerable effort has been devoted – and continues to be devoted – to the provision of basic health services at the village level. However, in a previous section, we reviewed the impact of accessibility on health, and found large discrepancies in outcomes on even the most basic health outcome indicators (infant mortality and life expectancy). The urban bias is simple to understand but often results in a widening gap between the poorer rural areas and the wealthier urban ones. Health is a sector where this bias is clear. Indonesia has made a very real effort to provide primary health centres to villages, and trained many thousands of doctors, nurses, primary health care workers, midwives and paramedics. Nevertheless, access to quality health care (even non-specialist) staff, pharmaceuticals and facilities is still very much an urban privilege.

However, the health budget is finite, and it is not realistic to expect that a comprehensive range of services can be universally available at the local level. Greater concentration of population and economic activity create an inbuilt bias in service provision towards metropolitan locations. The task, therefore, as Sumodiningrat says, is *'not to provide a hospital to every village ... but to make it* possible for rural communities to gain easier access to urbanbased facilities. This requires not only better physical infrastructures and transport systems but a rural health awareness campaign and means of paying for the required services'.

In other words, better transport is a necessary, though not a sufficient condition, of providing adequate access to health care for village communities. In the case of the remote, under-developed islands of Eastern Indonesia, where incomes for most remain well below the level that make air travel a realistic alternative, better transport means improved passenger shipping services.

Education

Analogous arguments can and have been advanced with respect to education services. While a key component in improving life prospects for the inhabitants of remote communities is the provision of sound basic education at the local level, it is unrealistic to expect the full range of educational opportunities and options to be available outside of major urban centres. Once again, therefore, reliable, efficient and affordable shipping services will play a key role in ensuring equitable access to educational opportunities for remote island residents.

Employment

Indonesia's strategy of diversifying its economic base to reduce its dependence on oil revenues has led to major changes in its economic structure, with a rapid rise in the importance of manufactured exports. Labour-intensive manufacturing increased particularly strongly during the latter half of the 1980's and early 1990's, creating a wide range of new employment opportunities.

As we have seen, these new opportunities are heavily concentrated in the main urban centres. One of consequences of this change has been a change in inter-provincial migration patterns. Indonesia has long had a transmigration programme to transfer people from the Inner (Java, Bali and Madura) to Outer Indonesia. Originally conceived as a programme to 'even out' population densities, the goals are now more commonly articulated in terms of development in the outer islands. Partly as a result of this programme, Java's population has been growing significantly more slowly than that of Indonesia as a whole.

However, Hugo notes that: The shift in Government policy in the late 1980s to facilitate international and domestic private investment and industrialisation is tending to favour growth in Java. Between 1985 and 1990 the number of people moving into Java (773,789) was almost as great as the number moving in the opposite direction.

Hugo further suggests that recently more people moved from the Outer Islands to Java than moved in the opposite direction'. As most inter-provincial migrants to Java settled in urban areas, the most probable explanation of this trend reversal is that migrants are 'attracted by the rapidly expanding urbanbased job opportunities'.

Official statistical data captures only permanent relocation. Perhaps even more important is the employment-induced temporary relocation of workers. It is widely accepted that the scale of non-permanent movements has increased dramatically in recent years, and is many times larger than permanent migration. Non-permanent migration increasingly provides an important source of supplementary income, and diversity of employment opportunities, to rural households. According to one study, 'twenty-five years ago many of the landless labourers on Java had very few sources of income...Now most of the landless rural families on Java have at least one person who is working outside the village, and in a factory or service job'.

Hugo cites a long list of reasons why temporary rather than permanent migration may be a preferred strategy for tapping the larger employment markets of the large cities. Some of these are causative, and some permissive. Amongst the most important are:

Causative:

- Participation in work in both the urban and rural sectors spreads the risk by diversifying families' portfolio of income-earning opportunities;
- The cost of living in urban areas is considerably higher, so that keeping the family in the village while earning in the city allows earnings to go further;
- Job options in the village, especially during seasonal increases in demand, are able to be kept open;
- In many cases, there is a preference for living and bringing up children in the village where there are seen to be fewer negative influences.

Permissive:

- Flexible time commitments in the urban informal sector allow time to circulate to the home village;
- The growing body of family and friends with urban experience makes the transition less intimidating for the rural worker, and often provides an urban base for him or her;
- Many urban employers provide barrack-style accommodation for workers;
- Recruiters and middlemen play a significantly increasing role in rural labour recruitment;
- Java's transport system is cheap and diverse, and allows workers to get to their home village from time to time.

All but the last of these considerations apply also to Outer Island residents. Migration to pursue employment opportunities is one of the most important mechanisms by which the 'trickle down' effect of industrial wealth generation is realised. The main focus of industrialisation in Indonesia will, for some time to come, be the major conurbation of Java, and it appears that temporary migration will be the preferred means of participation. For Outer Island residents and those in less well developed areas, the ability to take full advantage of these opportunities will require access to reliable and affordable transport services. Income levels suggest, and patronage statistics confirm, that these will be primarily maritime.

Previous Study on the Role of Transport for Poverty Alleviation

To examine the role of transport interventions in poverty alleviation, the United Nations ESCAP undertook a review of projects/

programmes in which transport could be a central element in alleviating poverty and improving the quality of life of people living in remote areas. The review comprised the following five case studies:

- the Rural Roads and Markets Improvement and Maintenance Project in Bangladesh;
- the Least-developed Village Development Grant Scheme in Indonesia;
- the Dhading Development Project and Gorkha Development Project in Nepal;
- the Aga Khan Rural Support Programme in Pakistan; and
- the Medium-term Development Plan in Philippines.

The case studies have demonstrated that poverty alleviation is a complex process, therefore, success or failure can rarely be attributed to one particular element within a programme. Nevertheless, transport interventions appear to have played a central role in the process of alleviating poverty or in improving the standard of living of the communities targeted in the respective projects mentioned above.

The followings summarize the major conclusions drawn from the case studies

with respect to the role of transport interventions in poverty alleviation.

- Transport interventions can be used as a policy instrument and an entry point for poverty alleviation.
- Transport interventions may have a direct impact on poverty reduction when the provision of improved transport is directly targeted towards the needs of the low-income groups and provides them with income earning opportunities. Direct impacts on the poor were observed in the cases of Indonesia and Philippines.
- Transport interventions also have an indirect social and welfare impact when improved transport provides cheaper and easier access to health, education and other services. Transport interventions in the cases of Bangladesh and Pakistan were found to have these indirect impacts on poverty alleviation.
- Transport interventions in Nepal had a combined impact generating employment opportunities as well as increasing social mobility.

Transportation Planning

Transportation planning is a field involved with the evaluation, assessment, design and siting of transportation facilities (generally streets, highways, footpaths, bike lanes and public transport lines).

Models and Sustainability

Transportation planning historically has followed the rational planning model of defining goals and objectives, identifying problems, generating alternatives, evaluating alternatives, and developing plans. Other models for planning include rational actor, transit oriented development, satisficing, incremental planning, organizational process, and political bargaining.

However, planners are increasingly expected to adopt a multi-disciplinary approach, especially due to the rising importance of environmentalism. For example, the use of behavioral psychology to persuade drivers to abandon their automobiles and use public transport instead. The role of the transport planner is shifting from technical analysis to promoting sustainability through integrated transport policies.

United Kingdom

In the United Kingdom transport planning has traditionally been a branch of civil engineering. In the 1950s and 1960s it was generally believed that the motor car was an important element in the future of transport as economic growth spurred on car ownership figures. The role of the transport planner was to match motorway and rural road capacity against the demands of economic growth. Urban areas would need to be redesigned for the motor vehicle or else impose traffic containment and demand management to mitigate congestion and environmental impacts. These policies were popularised in a 1963 government publication, Traffic in Towns. The contemporary Smeed Report on congestion pricing was initially promoted to manage demand but was deemed politically unacceptable. In more recent times this approach has been caricatured as "predict and provide" – to predict future transport demand and provide the network for it, usually by building more roads.

The publication of Planning Policy Guidance 13 in 1994 (revised in 2001), followed by A New Deal for Transport in 1998 and the white paper Transport Ten Year Plan 2000 again indicated an acceptance that unrestrained growth in road traffic was neither desirable nor feasible. The worries were threefold: concerns about congestion, concerns about the effect of road traffic on the environment (both natural and built) and concerns that an emphasis on road transport discriminates against vulnerable groups in society such as the poor, the elderly and the disabled.

These documents reiterated the emphasis on integration:

- integration within and between different modes of transport
- integration with the environment
- integration with land use planning
- integration with policies for education, health and wealth creation.

This attempt to reverse decades of underinvestment in the transport system has resulted in a severe shortage of transport planners. It was estimated in 2003 that 2,000 new planners would be required by 2010 to avoid jeopardising the success of the Transport Ten Year Plan.

During 2006 the Transport Planning Society defined the key purpose of transport planning as *to plan, design, deliver, manage and review transport, balancing the needs of society, the economy and the environment.*

The following key roles must be performed by transport planners:

- take account of the social, economic and environmental context of their work
- understand the legal, regulatory policy and resource framework within which they work
- understand and create transport policies, strategies and plans that contribute to meeting social, economic and environmental needs
- design the necessary transport projects, systems and services
- understand the commercial aspects of operating transport systems and services
- know about and apply the relevant tools and techniques
- must be competent in all aspects of management, in particular communications, personal skills and project management.

United States

Transportation planning in the United States is in the midst of a shift similar to that taking place in the United Kingdom, away from the singular goal of moving vehicular traffic and towards an approach that takes into consideration the communities and lands which streets, roads, and highways pass through ("the context").

More so, it places a greater emphasis on passenger rail networks which had been neglected until recently. This new approach, known as Context Sensitive Solutions (CSS), seeks to balance the need to move people efficiently and safely with other desirable outcomes, including historic preservation, environmental sustainability, and the creation of vital public spaces.

The initial guiding principles of CSS came out of the 1998 "Thinking Beyond the Pavement" conference as a means to describe and foster transportation projects that preserve and enhance the natural and built environments, as well as the economic and social assets of the neighbourhoods they pass through.

CSS principles have since been adopted as guidelines for highway design in federal legislation. And in 2003, the Federal Highway Administration announced that under one of its three Vital Few Objectives (Environmental Stewardship and Streamlining) they set the target of achieving CSS integration within all state Departments of Transportation by September of 2007. The recent pushes for advancing transportation planning has led to the development of a professional certification program, the Professional Transportation Planner, to be launched in 2007.

Transportation Forecasting

Transportation forecasting is the process of estimating the number of vehicles or travellers that will use a specific transportation facility in the future. A forecast estimates, for instance, the number of vehicles on a planned freeway or bridge, the ridership on a railway line, the number of passengers patronizing an airport, or the number of ships calling on a seaport. Traffic forecasting begins with the collection of data on current traffic. Together with data on population, employment, trip rates, travel costs, etc., traffic data are used to develop a traffic demand model. Feeding data on future population, employment, etc. into the model results in output for future traffic, typically estimated for each segment of the transportation infrastructure in question, e.g., each roadway segment or each railway station.

Traffic forecasts are used for several key purposes in transportation policy, planning, and engineering: to calculate the capacity of infrastructure, e.g., how many lanes a bridge should have; to estimate the financial and social viability of projects, e.g., using cost-benefit analysis and social impact assessment; and to calculate environmental impacts, e.g., air pollution and noise.

Four-step Models

Within the rational planning framework, transportation forecasts have traditionally followed the sequential four-step model or urban transportation planning (UTP) procedure, first implemented on mainframe computers in the 1950s at the Detroit Area Transportation Study and Chicago Area Transportation Study (CATS).

Land use forecasting sets the stage for the process. Typically, forecasts are made for the region as a whole, e.g., of population growth. Such forecasts provide control totals for the local land use analysis. Typically, the region is divided into zones and by trend or regression analysis, the population and employment are determined for each. The four steps of the classical urban transportation planning system model are:

- Trip generation determines the frequency of origins or destinations of trips in each zone by trip purpose, as a function of land uses and household demographics, and other socio-economic factors.
- Trip distribution matches origins with destinations, often using a gravity model function, equivalent to an entropy maximizing model. Older models include the fratar model.
- Mode choice computes the proportion of trips between each origin and destination that use a particular transportation mode. This model is often of the logit form, developed by Nobel Prize winner Daniel McFadden.
- Route assignment allocates trips between an origin and destination by a particular mode to a route. Often (for highway route assignment) Wardrop's principle of user equilibrium is applied (equivalent to a Nash equilibrium), wherein each traveller chooses the shortest (travel time) path, subject to every other driver doing the same. The difficulty is that travel times are a function of demand, while demand is a function of travel time, the so-called bi-level problem. Another approach is to use the Stackelberg competition model, where users ("followers") respond to the actions of a "leader", in this case for example a traffic manager. This leader anticipates on the response of the followers.

After the classical model, evaluative decision criteria are applied. A typical criterion is cost-benefit analysis. Such analysis might be applied after the network assignment model identifies needed capacity: is such capacity worthwhile? In addition to identifying the forecasting and decision steps as additional steps in the process, it is important to note that forecasting and decision-making permeate each step in the UTP process. Planning deals with the future, and it is forecasting dependent.

Activity-based Models

An example of an activity-based approach is the New York Metropolitan Transportation Council's Best Practice Model.

Precursor Steps

Although not identified as steps in the UTP process, a lot of data gathering is involved in the UTP analysis process. Census and land use data are obtained, and there are home interview surveys. Home interview surveys, land use data, and special trip attraction surveys provide the information on which the UTP analysis tools are exercised.

Data collection, management, and processing; model estimation; and use of models to yield plans are much used techniques in the UTP process. In the early days, census data was augmented that with data collection methods that had been developed by the Bureau of Public Roads (a predecessor of the Federal Highway Administration): traffic counting procedures, cordon "where are you coming from and where are you going" counts, and home interview techniques. Protocols for coding networks and the notion of analysis or traffic zones emerged at the CATS.

Model estimation used existing techniques, and plans were developed using whatever models had been developed in a study. The main difference between today and yesterday is the development of some analytic resources specific to transportation planning, in addition to the BPR data acquisition techniques used in the early days.

Critique

The sequential and aggregate nature of transportation forecasting has come under much criticism. While improvements have been made, in particular giving an activity-base to travel demand, much remains to be done. In the 1990s, most federal investment in model research went to the Transims project at Los Alamos National Laboratory, giving physicists a crack at the problem. While the use of supercomputers and the detailed simulations may be an improvement on practice, they have yet to be shown to be better (more accurate) than conventional models. A commercial version was spun off to IBM, and an open source version is also being actively maintained as TRANSIMS Open-Source. One of the major oversights in the use of transportation models in practice is the absence of any feedback from transportation models on land use. Highways and transit investments not only respond to land use, they shape it as well. There is growing interest in modeling land use and transportation interactions, by combining transportation models with land use models, such as UrbanSim.

Inaccuracy

Accurate traffic forecasts are critical to arriving at the right capacity for transportation infrastructure, that is, for building infrastructure that is neither too large or too small to meet the demand. Accurate traffic forecasts are also critical to obtaining valid results from the cost-benefit analyses, environmental impact assessments, and social impact assessment that typically form the basis for decisions on whether to build new transportation infrastructure or not.

To date there has been little research into this area. However, a peer-reviewed study of a large number of traffic forecasts found that a significant number of forecasts are inaccurate. In particular:

- for nine out of ten railway projects the study found that passenger forecasts were overestimated, with an average overestimate of 106%,
- for half of all road projects, including bridges and tunnels, the study found that the difference between actual and forecast traffic was more than 20%, while for 25% of road projects the difference was more than 40%.

Measured over decades, a scheme can have such a large turnover that even a small percentage change in the projected traffic can indicate a significant positive or negative economic effect. For road schemes, the study noted that changes in land use and difficulties in estimating journeys are often blamed for these errors. Sensitivity analysis is normally carried out to understand the performance of the scheme if the parameters change; but subsequent official policy decisions can also have a major effect, such as changes to the area plan, discount rates and values of time. For rail schemes, inaccuracies are often blamed on uncertain measurement of passenger journeys and deliberately slanted forecasts by over-optimistic promoters.

Vehicle Tracking System

A vehicle tracking system combines the installation of an electronic device in a vehicle, or fleet of vehicles, with purpose-designed computer software to enable the owner or a third party to track the vehicle's location, collecting data in the process. Modern vehicle tracking systems commonly use GPS or GLONASS technology for locating the vehicle, but other types of automatic vehicle location technology can also be used. Vehicle information can be viewed on electronic maps via the Internet or specialized software. Urban public transit authorities are an increasingly common user of vehicle tracking systems, particularly in large cities.

Active Versus Passive Tracking

Several types of Vehicle Tracking devices exist. Typically they are classified as "Passive" and "Active". "Passive" devices store GPS location, speed, heading and sometimes a trigger event such as key on/off, door open/closed. Once the vehicle returns to a predetermined point, the device is removed and the data downloaded to a computer for evaluation. Passive systems include auto download type that transfer data via wireless download. "Active" devices also collect the same information but usually transmit the data in real-time via cellular or satellite networks to a computer or data centre for evaluation. Many modern vehicle tracking devices combine both active and passive tracking abilities: when a cellular network is available and a tracking device is connected it transmits data to a server; when a network is not available the device stores data in internal memory and will transmit stored data to the server later when the network becomes available again.

Common Uses

Vehicle tracking systems are commonly used by fleet operators for fleet management functions such as routing, dispatch, on-board information and security. Along with commercial fleet operators, urban transit agencies use the technology for a number of purposes, including monitoring schedule adherence of buses in service, triggering changes of buses' destination sign displays at the end of the line (or other set location along a bus route), and triggering pre-recorded announcements for passengers. The American Public Transportation Association estimated that, at the beginning of 2009, around half of all transit buses in the United States were already using a GPS-based vehicle tracking system to trigger automated stop announcements.

This can refer to external announcements (triggered by the opening of the bus's door) at a bus stop, announcing the vehicle's route number and destination, primarily for the benefit of visually impaired customers, or to internal announcements (to passengers already on board) identifying the next stop, as the bus (or tram) approaches a stop, or both. Data collected as a transit vehicle follows its route is often continuously fed into a computer program which compares the vehicle's actual location and time with its schedule, and in turn produces a frequently updating display for the driver, telling him/her how early or late he/she is at any given time, potentially making it easier to adhere more closely to the published schedule. Such programs are also used to provide customers with real-time information as to the waiting time until arrival of the next bus or tram/streetcar at a given stop, based on the nearest vehicles' actual progress at the time, rather than merely giving information as to the scheduled time of the next arrival.

Transit systems providing this kind of information assign a unique number to each stop, and waiting passengers can obtain information by entering the stop number into an automated telephone system or an application on the transit system's website. Some transit agencies provide a virtual map on their website, with icons depicting the current locations of buses in service on each route, for customers' information, while others provide such information only to dispatchers or other employees.

Other applications include monitoring driving behaviour, such as an employer of an employee, or a parent with a teen driver. Vehicle tracking systems are also popular in consumer vehicles as a theft prevention and retrieval device. Police can simply follow the signal emitted by the tracking system and locate the stolen vehicle. When used as a security system, a Vehicle Tracking System may serve as either an addition to or replacement for a traditional Car alarm. Some vehicle tracking systems make it possible to control vehicle remotely, including block doors or engine in case of emergency. The existence of vehicle tracking device then can be used to reduce the insurance cost, because the loss-risk of the vehicle drops significantly.

Vehicle tracking systems are an integrated part of the "layered approach" to vehicle protection, recommended by the National Insurance Crime Bureau (NICB) to prevent motor vehicle theft. This approach recommends four layers of security based on the risk factors pertaining to a specific vehicle. Vehicle Tracking Systems are one such layer, and are described by the NICB as "very effective" in helping police recover stolen vehicles.

Some vehicle tracking systems integrate several security systems, for example by sending an automatic alert to a phone or email if an alarm is triggered or the vehicle is moved without authorization, or when it leaves or enters a geofence.

Other scenarios in which this technology is employed include:

- Stolen Vehicle Recovery: Both consumer and commercial vehicles can be outfitted with RF or GPS units to allow police to do tracking and recovery. In the case of LoJack, the police can activate the tracking unit in the vehicle directly and follow tracking signals.
- Fleet Management: When managing a fleet of vehicles, knowing the real-time location of all drivers allows management to meet customer needs more efficiently. Whether it is delivery, service or other multi-vehicle enterprises, drivers now only need a mobile phone with telephony or Internet connection to be inexpensively tracked by and dispatched efficiently.
- Asset Tracking: Companies needing to track valuable assets for insurance or other monitoring purposes can now plot the real-time asset location on a map and closely monitor movement and operating status.
- Field Service Management: Companies with a field service workforce for services such as repair or maintenance, must be able to plan field workers' time, schedule subsequent customer visits and be able to operate these departments efficiently. Vehicle tracking allows companies to quickly locate a field engineer and dispatch the closest one to meet a new customer request or provide site arrival information.
- Field Sales: Mobile sales professionals can access real-time locations. For example, in unfamiliar areas, they can locate themselves as well as customers and prospects, get driving directions and add nearby last-minute appointments to itineraries. Benefits include increased productivity, reduced driving time and increased time spent with customers and prospects.
- Trailer Tracking: Haulage and Logistics companies often operate lorries with detachable load carrying units. The part of the vehicle that drives the load is known as the cab and the load carrying unit is known as the trailer. There are different

types of trailer used for different applications, e.g., flat bed, refrigerated, curtain sider, box container.

- Surveillance: A tracker may be placed on a vehicle to follow the vehicle's movements.
- Transit Tracking: This is the temporary tracking of assets or cargoes from one point to another. Users will ensure that the assets do not stop on route or do a U-Turn in order to ensure the security of the assets.

Vehicle Tracking Systems are widely used worldwide. Components come in various shapes and forms but most utilize GPS technology and SMS Messaging. While most will offer real-time tracking, Others record real time data and store it to be read, similar to data-loggers. Systems like these track and record and allow reports after certain points have been saved.

Vehicle Infrastructure Integration

Vehicle Infrastructure Integration (VII) is an initiative fostering research and applications development for a series of technologies directly linking road vehicles to their physical surroundings, first and foremost in order to improve road safety. The technology draws on several disciplines, including transport engineering, electrical engineering, automotive engineering, and computer science. VII specifically covers road transport although similar technologies are in place or under development for other modes of transport. Planes, for example, use ground-based beacons for automated guidance, allowing the autopilot to fly the plane without human intervention. In highway engineering, improving the safety of a roadway can enhance overall efficiency. VII targets improvements in both safety and efficiency. Vehicle infrastructure integration is that branch of engineering, which deals with the study and application of a series of techniques directly linking road vehicles to their physical surroundings in order to improve road safety.

Goals

The goal of VII is to provide a communications link between vehicles on the road (via On-Board Equipment, OBE), and between vehicles and the roadside infrastructure (via Roadside Equipment, RSE), in order to increase the safety, efficiency, and convenience of the transportation system. It is based on widespread deployment of a dedicated short-range communications (DSRC) link, incorporating IEEE 802.11p. VII's development relies on a business model supporting the interests of all parties concerned: industry, transportation authorities and professional organisations. The initiative has three priorities:

- evaluation of the business model (including deployment scheduling) and acceptance by the stakeholders;
- validation of the technology (in particular the communications systems) in the light of deployment costs; and
- development of legal structures and policies (particularly in regard to privacy) to enhance the system's potential for success over the longer term.

Safety

Current active safety technology relies on vehicle-based radar and vision systems. For example, this technology can reduce rear-end collisions by tracking obstructions in front or behind the vehicle, automatically applying brakes when needed. This technology is somewhat limited in that it senses only the distance and speed of vehicles within the direct line of sight. It is almost completely ineffective for angled and left-turn collisions. It may even cause a motorist to lose control of the vehicle in the event of an impending head-on collision. The rear-end collisions covered by today's technology are typically less severe than angle, left-turn, or head-on collisions. Existing technology is therefore inadequate for the overall needs of the roadway system.

VII would provide a direct link between a vehicle on the road and all vehicles within a defined vicinity. The vehicles would be able to communicate with each other, exchanging data on speed, orientation, perhaps even on driver awareness and intent. This could increase safety for nearby vehicles, while enhancing the overall sensitivity of the VII system, for example, by performing an automated emergency manoeuvre (steering, decelerating, braking) more effectively. In addition, the system is designed to communicate with the roadway infrastructure, allowing for complete, real-time traffic information for the entire network, as well as better queue management and feedback to vehicles. It would ultimately close the feedback loops on what is now an open-loop transportation system.

Through VII, roadway markings and road signs could become obsolete. Existing VII applications use sensors within vehicles which can identify markings on the roadway or signing along the side of the road, automatically adjusting vehicle parameters as necessary.

Ultimately, VII aims to treat such signs and markings as little more than stored data within the system. This could be in the form of data acquired via beacons along a roadway or stored at a centralised database and distributed to all VII-equipped vehicles.

Efficiency

All the above factors are largely in response to safety but VII could lead to noticeable gains in the operational efficiency of a transportation network. As vehicles will be linked together with a resulting decrease in reaction times, the headway between vehicles could be reduced so that there is less empty space on the road. Available capacity for traffic would therefore be increased. More capacity per lane will in turn mean fewer lanes in general, possibly satisfying the community's concerns about the impact of roadway widening. VII will enable precise traffic-signal coordination by tracking vehicle platoons and will benefit from accurate timing by drawing on real-time traffic data covering volume, density and turning movements.

Real-time traffic data can also be used in the design of new roadways or modification of existing systems as the data could be used to provide accurate origin-destination studies and turning-movement counts for uses in transportation forecasting and traffic operations. Such technology would also lead to improvements for transport engineers to address problems whilst reducing the cost of obtaining and compiling data. Tolling is another prospect for VII technology as it could enable roadways to be automatically tolled. Data could be collectively transmitted to road users for in-vehicle display, outlining the lowest cost, shortest distance, and/or fastest route to a destination on the basis of real-time conditions.

Existing Applications

To some extent, results along these lines have been achieved in trials performed in around the globe, making use of GPS, mobile phone signals, and vehicle registration plates. GPS is becoming standard in many new high-end vehicles and is an option on most new low-and mid-range vehicles. In addition, many users also have mobile phones which transmit trackable signals (and may also be GPS-enabled). Mobile phones can already be traced for purposes of emergency response. GPS and mobile phone tracking, however, do not provide fully reliable data. Furthermore, integrating mobile phones in vehicles may be prohibitively difficult. Data from mobile phones, though useful, might even increase risks to motorists as they tend to look at their phones rather than concentrate on their driving. Automatic registration plate recognition can provide high levels of data, but continuously tracking a vehicle through a corridor is a difficult task with existing technology. Today's equipment is designed for data acquisition and functions such as enforcement and tolling, not for returning data to vehicles or motorists for response. GPS will nevertheless be one of the key components in VII systems.

Limitations

There are numerous limitations to the development of VII. A common misconception is that the biggest challenge to VII technology is the computing power that can be fitted inside a vehicle. While this is indeed a challenge, the technology for computers has been advancing rapidly and is not a particular concern for VII researchers. Given the fact that technologies already exist for the most basic of forms of VII, perhaps the greatest hurdle to the deployment of VII technology is public acceptance.

Privacy

The most common myth about VII is that it includes tracking technology; however, this is not the case. The architecture is designed to prevent identification of individual vehicles, with all data exchange between the vehicle and the system occurring anonymously. Exchanges between the vehicles and third parties such as OEMs and toll collectors will occur, but the network traffic will be sent via encrypted tunnels and will therefore not be decipherable by the VII system.

Although the system will be able to detect signal and speed violations, it will not have the capability to identify the violator and report them. The detection is for the purpose of alerting the violator and/or approaching vehicles, to prevent collisions.

Other Public Concerns

Other public acceptance concerns come from advocates of recreational driving as well as from critics of tolling. The former argue that VII will increase the automation of the vehicle, reducing the driver's enjoyment. Recreational driving concerns are particularly prevalent among owners of sports cars. They could be attenuated by compensating for the presence of vehicles without VII or perhaps by maintaining roadways where vehicles without VII are permitted to travel.

Those opposed to tolling believe it will make driving prohibitively expensive for motorists in the lower-income bracket, conflicting with the general wish to provide equal services for all. In response, public transit discounts or road use discounts can be considered for qualifying individuals and/or families. Such provisions currently exist for numerous tolled roadways and could be applicable to roadways that are tolled via VII. However, as VII could allow for the tolling of *every* VII-enabled roadway, the provisions may be ineffective in view of the increased need to provide user-efficient transit services to every area.

Technical Issues

Coordination

A major issue facing the deployment of VII is the problem of how to stand up the system initially. The costs associated with installing the technology in vehicles and providing communications and power at every intersection are significant. Building out the infrastructure along the roadside without the auto manufacturers' cooperation would be disastrous, as would the reverse situation; therefore, the two parties will need to work together to make the VII concept work.

There are proof of concept tests being performed in Michigan and California that will be evaluated by the US DOT and the auto manufacturers, and a decision will be made, jointly, about whether or not to move forward with implementation of the system at that time.

Maintenance

Another factor for consideration in regard to the technology's distribution is how to update and maintain the units. Traffic systems are highly dynamic, with new traffic controls implemented every day and roadways constructed or repaired every year. The vehicle-based option could be updated via the internet (preferably wireless), but may subsequently require all users to have access to internet technology. Many local government agencies have been testing deployment of internet facilities in cities and along roadways, for example at rest-stops. These systems could be used for VII updating.

An additional option is to provide updates whenever a vehicle is brought in for inspection or servicing. A major limitation here is that updating would be in the hands of the user. Some vehicle owners maintain their vehicles themselves, and periodic inspections or servicing are considered too infrequent for updating VII. Motorists might also be reluctant to stop at rest-stops for an update if they do not have the possibility of driving in an internet-enabled city.

Alternatively, if receivers were placed in all vehicles and the VII system was primarily located along the roadside, information could be stored in a centralised database. This would allow the agency responsible to issue updates at any time. These would then be disseminated to the roadside units for passing motorists. Operationally, this method is currently considered to provide the greatest effectiveness but at a high cost to the authorities.

Security

Security of the units is another concern, especially in the light of the public acceptance issue. Criminals could tamper with VII units, or remove and/or destroy them regardless of whether they are installed inside vehicles or along the roadside. If they are placed inside vehicles, laws similar to those for tampering with an odometer could be enacted; and the units could be examined during inspections or services for signs of tampering. This method has many of the limitations mentioned in relation to the frequency of inspection and motorists who perform their own servicing. It also raises concerns regarding the honesty of vehicle technicians performing the inspections. The ability of technicians to identify signs of tampering would be dependent on their knowledge of the VII systems themselves.

Magnets, electric shocks, and malicious software (viruses, hacking, or jamming) could be used to damage VII systems-regardless of whether units are located inside vehicle or along the roadside. Extensive training and certification would be required for technicians to inspect VII units within a vehicle. Along the roadside, a high degree of security would be required to ensure that the equipment is not damaged and to increase its durability. However, as roadside units could well be placed on the public right-of-way-which is often close to the edge of the roadway-there could be concerns about vehicles hitting them (whether on purpose or by accident). The units would either have to be built so that they do not provide a threat to motorists: perhaps in the form of a low-profile and/or low-mass object designed to be run over or to break apart (which would entail a relatively inexpensive unit); or the unit would have to be shielded by a device such as a guardrail, raising safety concerns of its own.

Data Input

Yet another limitation is in digitizing the inputs for the VII system. VII systems will probably continue to sense existing signs and roadway markings but one of the goals is to eliminate such signs and markings altogether. This would require converting the locations and messages of each item into the VII system's format. Responsibility for this work would probably fall on the highway agencies which nearly all face difficulties in funding, manpower, and available time. Implementing and maintaining VII systems may therefore require support at the national level.

Communications and Authorization

While VII is largely being developed as a joint research enterprise involving numerous transport agencies, it is likely initial products will be tailored to individual applications. As a result, compatibility and formatting issues could well arise as systems expand. Overcoming these difficulties could require complicated translation programs between different systems or possibly a complete overhaul of existing VII systems in order to develop a more comprehensive approach. In either case, the costs and potential for bugs in the software will likely be high.

Legislation will be required to set in place access to the VII data and communications between applicable agencies. In the USA, for example, an Interstate is a Federal roadway that is often maintained by the State, but the local county or municipal authorities may be involved too. The legislation would need to set the levels of authority of each agency. In Pennsylvania, for example, municipalities tend to have greater authority than counties and sometimes even the State whereas neighboring Maryland has more authority at the county level than at municipal level; and State roads are almost exclusively controlled by the State. It would also have to be determined which other agencies can use the data (i.e. law enforcement, Census, etc.) and to what degree it is permissible to use the information. Law enforcement would be needed to minimise data misuse. The various levels of authority could also increase incompatibility.

Recent Developments

Much of the current research and experimentation is conducted in the United States where coordination is ensured through the Vehicle Infrastructure Integration Consortium, consisting of automobile manufacturers (Ford, General Motors, DaimlerChrysler, Toyota, Nissan, Honda, Volkswagen, BMW), IT suppliers, U.S. Federal and state transportation departments, and professional associations. Trialling is taking place in Michigan and California.

The specific applications now being developed under the U.S. initiative are:

- Warning drivers of unsafe conditions or imminent collisions.
- Warning drivers if they are about to run off the road or speed around a curve too fast.
- Informing system operators of real-time congestion, weather conditions and incidents.
- Providing operators with information on corridor capacity for real-time management, planning and provision of corridor-wide advisories to drivers.

In mid-2007, a VII environment covering some 20 square miles (52 km) near Detroit will be used to test 20 prototype VII applications. Several automobile manufacturers are also conducting their own VII research and trialling.

8

Transport in India

Transport in the Republic of India is an important part of the nation's economy. Since the economic liberalisation of the 1990s, development of infrastructure within the country has progressed at a rapid pace, and today there is a wide variety of modes of transport by land, water and air. However, the relatively low GDP of India has meant that access to these modes of transport has not been uniform. Motor vehicle penetration is low with only 13 million cars on the nation's roads. In addition, only around 10% of Indian households own a motorcycle. At the same time, the Automobile industry in India is rapidly growing with an annual production of over 2.6 million vehicles and vehicle volume is expected to rise greatly in the future. In the interim however, public transport still remains the primary mode of transport for most of the population, and India's public transport systems are among the most heavily utilised in the world. India's rail network is the longest and fourth most heavily used system in the world transporting over 6 billion passengers and over 350 million tons of freight annually.

Despite ongoing improvements in the sector, several aspects of the transport sector are still riddled with problems due to outdated infrastructure, lack of investment, corruption and a burgeoning population. The demand for transport infrastructure and services has been rising by around 10% a year with the current infrastructure being unable to meet these growing demands. According to recent estimates by Goldman Sachs, India will need to spend $1.7 Trillion USD on infrastructure projects over the next decade to boost economic growth of which $500 Billion USD is budgeted to be spent during the eleventh Five-year plan.

Traditional Means

Walking

In ancient times, people often covered long distances on foot. For instance, Adi Sankaracharya traveled all over India. Walking still constitutes an important mode of transport in urban areas. In the city of Mumbai, to further improve the transit conditions for pedestrians, the Mumbai Metropolitan Region Development Authority, has commenced the construction of more than 50 skywalks, as part of the Mumbai Skywalk project.

Palanquin

Palanquins also known as *palkis*, were one of the luxurious methods used by the rich and noblemen for travelling. This was primarily used in the olden days to carry a deity or idol of a god, and many temples have sculptures of god being carried in a *palki*. Later on, it was primarily used by European noblemen and ladies from the upper classes of society prior to the advent of the railways in India. Modern use of the palanquin is limited to being an ostentatious method for the bride to enter Indian weddings.

Bullock Carts and Horse Carriages

Bullock carts have been traditionally used for transport, especially in rural India. The advent of the British saw drastic improvements in the horse carriages which were used for transport since early days. Today, they are used in smaller towns and are referred as Tonga or *buggies*. Victorias of Mumbai are still used for tourist purposes, but horse carriages are now rarely found in the metro cities of India. In recent years some cities have banned the movement of bullock carts and other slow moving vehicles on the main roads.

Bicycles

Bicycles are a common mode of travel in much of India. More people can now afford to own a cycle than ever before. In 2005, more than 40% of Indian households owned a bicycle, with ownership rates ranging from around 30% to 70% at the state level. Along with walking, cycling accounts for 50 to 75 % of the commuter trips for those in the informal sector in urban areas.

Even though India is the second largest producer of bicycles in the world, a significant prejudice against bicycle riding for transport exists in some segments of the population, generally stemming from the status symbol aspect of the motor vehicle. In India, the word "bike" generally refers to motorcycle, and "cycle" refers to bicycle. Pune was the first city in India to have dedicated lanes for cycles. It was built for the 2008 Commonwealth Youth Games.

However, recent developments in Delhi suggest that bicycle riding is fast becoming popular in the metro cities of India. The Delhi government has decided to construct separate bicycle lanes on all major roads to combat pollution and ease traffic congestion.

Hand-pulled Rickshaw

This type of transport is still available in Kolkata wherein a person pulls the rickshaw by hand. The Government of West Bengal proposed a ban on these rickshaws in 2005 describing them as "inhuman". Though a bill aiming to address this issue, termed as 'Calcutta Hackney Carriage Bill', was passed by the West Bengal Assembly in 2006, it has not been implemented yet. The Government of West Bengal is working on an amendment of this bill to avoid the loopholes that got exposed when the Hand-pulled Rickshaw Owner's Association filed a petition against the bill.

Cycle Rickshaw

Cycle rickshaws were introduced into India in the 1940s. They are bigger than a tricycle where two people sit on an elevated seat at the back and a person pedals from the front. In the late 2000s, they were banned in several cities for causing traffic congestion. Cycle rickshaws have been a feature of Delhi streets since Indian independence in 1947, providing the cheapest way around the capital. The Delhi Police recently submitted an affidavit against plying of cycle rickshaws to ease traffic congestion in the city but it was dismissed by the Delhi High court. In addition, environmentalists have supported the retention of cycle rickshaws as a non-polluting and inexpensive mode of transport.

Trams

The advent of the British saw trams being introduced in many cities including Mumbai and Kolkata. They are still in use in Kolkata and provide an emission-free means of transport. The nationalized Calcutta Tramways Company is in the process of upgrading the existing tramway network at a cost of Rs. 240 million.

Kolkata

The Calcutta Tramways Company Limited is the company which manages tramways in Kolkata. The Tram Service Was Started In

CALCUTTA For The First Time. The first horse driven tram carriage service started in the year 1873 between Sealdah and Armenian Ghat Street (Inaugurated on February 24, 1873). The service was wound up on November 20 as there was no adequate patronisation. The British started The Calcutta Tramways Company Limited as a registered joint stock company at London, in the year 1880. Before 1900s they were powered by horses later the same year the process of electrification began. In the year 1951 The Government of West Bengal entered into an agreement with the Calcutta Tramways Co and the Calcutta Tramways Act, 1951 was enacted. The Government took over all rights with regard to Tramways and reserved the right to purchase the system on 1st Jan, 1972 or any time thereafter giving two years notice. Later in the year 1967 The Govt of West Bengal passed the Calcutta Tramways Company (Taking Over of Management) Act, 1967 and took over the management on 19 July 1967. On November 8, 1976 the Calcutta Tramways (Acquisition of Undertaking) Ordinance, 1976 was promulgated under which the Company with all its assets vested with the government. After that it is now a Public Sector Undertaking.

Mumbai

The British proposed the introduction of trams in the year 1864 and the contract was awarded to Stearns and Kitteredge in the year 1873. The first tram was between Parel and Colaba were drawn by six to eight horses. The trams had their share of glory till the advent of trains and were there for almost 21 years and it is also said that the Stearns and Kitteredge had a stable of 900 horses when they started the service.

Patna

Patna was among pioneer selected towns of India having horse-drawn trams as urban transport. The horse drawn tram in Patna drove in the populated stretch of Ashok Rajpath, from Patna City to Bankipore, with its western terminal point at Sabzibagh (opposite Pirbahore Police Station) under the charge of the Patna City Municipality. The tram was discontinued in 1903, due to non-viability. Though the government had planning to stretch it further west, it did not materialise.

Kanpur

Trams were introduced in Kanpur in Jun 1907. The introductory stocks were single coach in Kanpur like other Indian cities (Delhi,

Mumbai and Chennai), because the new mode of transport was introduced experimentally. They were electric Traction Type.

The anti-tram craze started around 1955, and quickly spread up around the world. Many countries of both developed and developing countries started closing tram systems. India was not the exception. Tram service gradually closed at Kanpur in 16 May 1933.

There were 4 miles of track and 20 single-deck open trams. The single line connected the railway station with Sirsaya Ghat on the banks of the Ganges. Photographs of Cawnpore trams are very rare.

Nasik

This tramway was constructed in 1889 to a gauge of 2 ft 6 in (762 mm). Consulting engineer was Everard Calthrop, who later achieved fame with the Barsi Light Railway. Originally the tramway used two carriages pulled by four horses, but petrol driven railcars were introduced in 1916.

It originated from what is now the "Old municipal corporation" building located on "main road" and terminated at Nasik Road Railway station, a distance of around 8-10 km. The strip between Nasik and Nasik Road was covered with dense jungle and only mode of transport from the station to the city was by means of horse carriages or by means of only 2 taxis which used to ply in those days. The tramway closed down around 1931-1933.

Chennai

Trams in Chennai (old Madras) were operated between ports and other places. This could carry lot of head loads and was a very popular transport at that time. This ended by about 1950, when the tramway company went into liquidation. The contract to remove the tracks and overhead cables was awarded to one Mr.Narainsingh Ghanshamsingh. This is also shown in the film Madarasapatnam.

Local Transport

Public transport is the predominant mode of motorised local travel in cities. This is predominantly by road, since commuter rail services are available only in the four metropolitan cities of Mumbai, Delhi, Chennai, and Kolkata, while dedicated city bus services are known to operate in at least 17 cities with a population of over one million. Intermediate public transport modes like tempos and cycle rickshaws assume importance in medium size cities. However, the share of buses is negligible in most Indian cities as compared to personalized vehicles, and two-wheelers and cars account for more than 80 percent of the vehicle population in most large cities.

Traffic in Indian cities generally moves slowly, where traffic jams and accidents are very common. India has very poor records on road safely—around 90,000 people die from road accidents every year. A Reader's Digest study of traffic congestion in Asian cities ranked several Indian cities within the Top Ten for worst traffic.

Public Transport

Buses

Buses take up over 90% of public transport in Indian cities, and serve as a cheap and convenient mode of transport for all classes of society. Services are mostly run by state government owned transport corporations. However, after the economic liberalisation, many state transport corporations have introduced various facilities like low-floor buses for the disabled and air-conditioned buses to attract private car owners to help decongest roads. Bengaluru was the first city in India to introduce Volvo B7RLE intra-city buses in India in January 2006.

New initiatives like Bus Rapid Transit (BRT) systems and air conditioned buses have been taken by the various state government to improve the bus public transport systems in cities. Bus Rapid Transit systems already exist in Pune, Delhi and Ahmedabad with new ones coming up in Vishakhapatnam and Hyderabad. High Capacity buses can be found in cities like Mumbai, Bengaluru, Nagpur and Chennai. Bengaluru is the first Indian city to have an air-conditioned bus stop, located near Cubbon Park. It was built by Airtel. The city of Chennai houses Asia's largest bus terminus, the Chennai Mofussil Bus Terminus. In 2009, the Government of Karnataka and the Bangalore Metropolitan Transport Corporation flagged off a pro-poor bus service called the Atal Sarige. The service aims to provide low-cost connectivity to the economically backward sections of the society to the nearest major bus station.

Auto Rickshaws

An auto rickshaw is a three wheeler vehicle for hire that has no doors and is generally characterised by a small cabin for the driver in the front and a seat for passengers in the rear. Generally it is painted in yellow, green or black colour and has a black, yellow or green canopy on the top, but designs vary considerably from place to place.

In Mumbai and other metropolitan cities, 'autos' or 'ricks' as they are popularly known have regulated metered fares. A recent law prohibits auto rickshaw drivers from charging more than the specified fare, or charging night-fare before midnight, and also prohibits the driver from refusing to go to a particular location. Mumbai is also the only city which prohibits these vehicles from entering a certain part of the city, in this case being South Mumbai. In Chennai, it is common to see autorickshaw drivers demand more than the specified fare and refuse to use fare meter

Airports and railway stations at many cities such as Bengaluru, and Hubballi-Dharwad provide a facility of prepaid auto booths, where the passenger pays a fixed fare as set by the authorities for various locations.

Taxi

Most of the traditional taxicabs in India are either Premier Padmini or Hindustan Ambassador cars. In recent years, cars such as Chevrolet Tavera, Maruti Esteem, Maruti Omni, Mahindra Logan, Tata Indica, Toyota Innova and Tata Indigo have become fairly popular among taxi operators.

The livery of the taxis in India varies from state-to-state. In Delhi and Maharashtra, most taxicabs have yellow-black livery while in West Bengal, taxis have yellow livery. Private taxi operators are not required to have a specific livery. However, they are required by law to be registered as commercial vehicles.

Depending on the city/state, taxis can either be hailed or hired from taxi-stands. In cities such as Bengaluru, Hyderabad,taxis need to be hired over phone, whereas in cities like Kolkata and Mumbai, taxis can be hailed on the street. According to government of India regulations, all taxis are required to have a fare-meter installed. There are additional surcharges for luggage, late-night rides and toll taxes are to be paid by the passenger. Since 2006, radio taxis have become increasingly popular with the public due to reasons of safety and convenience.

In cities and localities where taxis are expensive or do not ply as per the government or municipal regulated fares, people use share taxis. These are normal taxis which carry one or more passengers travelling to destinations either en route to the final destination, or near the final destination. The passengers are charged according to the number of people with different destinations. A similar system exists for autorickshaws, known as share autos.

The city of Mumbai will soon be the first city in India, to have an "in-taxi" magazine, titled MumBaee, which will be issued to taxis which are part of the Mumbai Taximen's Union. The magazine is set to debut on the 13 July 2009.

Rail

The present suburban railway services in India are extremely limited and are operational only in Mumbai, Kolkata, Chennai and Delhi. The Mumbai Suburban Railway which began services in Mumbai in 1867, transports 6.3 million passengers daily and has the highest passenger density in the world.

The first rapid transit system in India, the Kolkata Suburban Railway, was established in Kolkata in 1854. Its first service ran between Howrah and Hooghly covering a distance of 38.6 km (24 mi). The Delhi Metro followed in 2002 and has carried over a billion commuters in seven years since its inauguration. Apart from these, Kolkata has a circular rail line and Chennai has an elevated rail transit called MRTS. Kolkata was the first city in India to possess a subterreanean rapid transport system, the *Kolkata Metro*, whose operations commenced in 1984. Rapid transit systems are also under construction in Hyderabad, Bengaluru, Chennai, Ahmedabad and Mumbai.

Rapid transit systems have been proposed in Thane, Pune, Kanpur, Lucknow, Amritsar and Kochi. Mumbai will soon be one of the two cities in India to have a monorail network, which is presently under construction. There are also monorail systems being planned in Kolkata and in Delhi. The Konkan Railway Corporation had patented a suspended monorail system called the Skybus Metro in Margao, but this is yet to be implemented anywhere on a commercial scale following an accident in 2004. A two-track elevated corridor has been proposed above the existing Western Railway line between the stations of Churchgate and Virar in Mumbai for air-conditioned EMUs.

Two-wheelers

Motorised two-wheelers like scooters, small capacity motorcycles and mopeds are very popular as a mode of transport due to their fuel efficiency and ease of use in congested traffic. The number of two-wheelers sold is several times that of cars. There were 4.75 crore (47.5 million) powered two wheelers in India in 2003 compared with just 86 lakh (8.6 million) cars. Hero Honda, Honda, TVS Motors and Bajaj Auto are the largest two-wheeler companies in terms of market-share.

Royal Enfield, an iconic brand name in the country, manufactures different variants of the Bullet motorcycle which is regarded as a classic motorcycle that is still in production.

Manufacture of scooters in India started when *Automobile Products of India (API)*, set up at Mumbai and incorporated in 1949, began assembling Innocenti-built Lambretta scooters in India post independence. They eventually acquired licence for the Li150 series model, of which they began full-fledged production from the early sixties onwards. In 1972, *Scooters India Ltd (SIL)*, a state-run enterprise based in Lucknow, Uttar Pradesh, bought the entire manufacturing rights of the last Innocenti Lambretta model. API has infrastructural facilities at Mumbai, Aurangabad, and Chennai but has been non-operational since 2002. SIL stopped producing scooters in 1998.

Motorcycles and scooters can be rented in many cities. Wearing protective headgear is mandatory for both the rider and the pillion-rider in most cities.

Automobiles

Private vehicles account for 30% of the total transport demand in urban areas of India. An average of 963 new private vehicles are registered every day in Delhi alone. The number of automobiles produced in India rose from 63 lakh (6.3 million) in 2002-03 to 1.1 crore (11.2 million) in 2008-09. However, India still has a very low rate of car ownership. When comparing car ownership between BRIC developing countries, it is on a par with China, and exceeded by Brazil and Russia. Compact cars, especially hatchbacks predominate due to affordability, fuel efficiency, congestion, and lack of parking space in most cities. Maruti, Hyundai and Tata Motors are the most popular brands in the order of their market share.

The Ambassador once had a monopoly but is now an icon of pre-liberalisation India, and is still used by taxi companies. Maruti 800 launched in 1984 created the first revolution in the Indian auto sector because of its low pricing. It had the highest market share until 2004, when it was overtaken by other low cost models from Maruti such as the Alto and the Wagon R, the Indica from Tata Motors and the Santro from Hyundai. Over the 20 year period since its introduction, about 24 lakh (2.4 million) units of the Maruti 800 have been sold. However, with the launch of the Tata Nano, the least expensive production car in the world, this is under threat.

India is also known for a variety of indigenous vehicles made in villages out of simple motors and vehicle spare-parts. A few of these innovations are the Jugaad, *Maruta*, *Chhakda*, and the *Fame*.

In the city of Bengaluru, Radio One and the Bangalore Traffic Police, launched a carpooling drive which is has involved celebrities such as Robin Uthappa, and Rahul Dravid encouraging the public to carpool. The initiative got a good response, and by the end of May 2009, 10,000 people are said to have carpooled in the city.

Utility Vehicles

The first utility vehicle in India was manufactured by Mahindra and Mahindra. It was a copy of the original Jeep and was manufactured under licence. The vehicle was an instant hit and made Mahindra one of the top companies in India. The Indian Army and police extensively use Mahindra vehicles along with Maruti Gypsys for transporting personnel and equipment.

Tata Motors, the automobile manufacturing arm of the Tata Group, launched its first utility vehicle, the Tata Sumo, in 1994. The Sumo, owing to its then-modern design, captured a 31% share of the market within two years. The Tempo trax from Force Motors till recently was ruling the rural areas. Sports utility vehicles now form a sizeable part of the passenger vehicle market. Models from Tata, Honda, Hyundai, Ford, Chevrolet and other brands are available.

Long Distance Transport

Railway

Rail services in India, first introduced in 1853, are provided by the state-run Indian Railways, under the supervision of the Ministry of Railways. Indian Railways provides an important mode of transport in India, transporting over 18 million passengers and more than 2 million tonnes of freight daily across one of the largest and busiest rail networks in the world. By 1947, the year of India's independence, there were forty-two rail systems. In 1951 the systems were nationalised as one unit, becoming one of the largest networks in the world. Indian Railways is divided into sixteen zones, which are further sub-divided into sixty seven divisions, each having a divisional headquarters. The rail network traverses through the length and breadth of the country, covering 6,909 stations over a total route length of around 63,465 km (39,435 mi). It is the world's largest commercial or utility employer, with more than 1.4 million employees. As to rolling stock, IR owns over 200,000 (freight) wagons, 50,000 coaches and 8,000 locomotives. It also owns locomotive and coach production facilities. It operates both long distance and suburban rail systems on a multi-gauge network of broad, metre and narrow gauges, and is in the process of converting all the metre gauge (14,406 km (8,951 mi)) into broad gauge in a project called Project Unigauge.

Kashmir Railway is the second highest in the world and the first phase was completed in 2009. Proposals have been made to introduce high-speed rail in India. A proposal has been made to build a Maglev track within the city of Mumbai, connecting it to the National Capital of New Delhi, as well as other parts of Maharashtra in the form of the Mumbai Maglev. Another proposal has been made to introduce a High-speed rail in India similar to that of the Shinkansen of Japan.

In 1999, the Konkan Railway Corporation introduced the *Roll On Roll Off* (RORO) service, a unique road-rail synergy system, on the section between Kolad in Maharashtra and Verna in Goa, which was extended up to Surathkal in Karnataka in 2004. The RORO service, the first of its kind in India, allowed trucks to be transported on flatbed trailers. It was highly popular, carrying about 1,10,000 trucks and bringing in about Rs.74 crore worth of earnings to the corporation till 2007.

Indian Railways

Indian Railways, abbreviated as IR, is the state-owned railway company of India, which owns and operates most of the country's rail transport. It is overseen by the Ministry of Railways of the Government of India.

Indian Railways has the largest rail network in Asia and the world's second largest under one management, transporting 20 million passengers and more than 2 million tonnes of freight daily. It is one of the world's largest commercial or utility employers, with more than 1.6 million employees. The railways traverse the length and breadth of the country, covering 6,909 stations over a total route length of more than 63,327 kilometres (39,350 mi). As to rolling stock, IR owns over 200,000 (freight) wagons, 50,000 coaches and 8,000 locomotives.

Railways were first introduced to India in 1853. By 1947, the year of India's independence, there were forty-two rail systems. In 1951 the systems were nationalised as one unit, becoming one of the largest networks in the world. IR operates both long distance and suburban rail systems on a multi-gauge network of broad, metre and narrow gauges. It also owns locomotive and coach production facilities. Initially, the Indian railways were both designed and built by the British, during their colonial rule of the subcontinent.

Organisational Structure

Indian Railways is a department owned and controlled by the Government of India, via the Ministry of Railways. As of May 2010, the Railway Ministry is headed by Mamata Banerjee, the Union Minister for Railways, and assisted by two ministers of State for Railways. Indian Railways is administered by the Railway Board, which has a financial commissioner, five members and a chairman.

Railway Zones

Indian Railways is divided into zones, which are further sub-divided into divisions. The number of zones in Indian Railways increased from six to eight in 1951, nine in 1952, and finally 16 in 2003. Each zonal railway is made up of a certain number of divisions, each having a divisional headquarters. There are a total of sixty-seven divisions.

The Kolkata Metro is owned and operated by Indian Railways, but is not a part of any of the zones. It is administratively considered to have the status of a zonal railway.

Each of the sixteen zones, as well as the Kolkata Metro, is headed by a General Manager (GM) who reports directly to the Railway Board. The zones are further divided into divisions under the control of Divisional Railway Managers (DRM). The divisional officers of engineering, mechanical, electrical, signal and telecommunication, accounts, personnel, operating, commercial and safety branches report to the respective Divisional Manager and are in charge of operation and maintenance of assets. Further down the hierarchy tree are the Station Masters who control individual stations and the train movement through the track territory under their stations' administration.

Recruitment and Training

With approximately 1.6 million employees, Indian Railways is the country's single largest employer. Staff are classified into gazetted (Group A and B) and non-gazetted (Group C and D) employees.

The recruitment of Group A gazetted employees is carried out by the Union Public Service Commission through exams conducted by it. The recruitment to Group 'C' and 'D' employees on the Indian Railways is done through 19 Railway Recruitment Boards which are controlled by the Railway Recruitment Control Board (RRCB). The training of all cadres is entrusted and shared between six centralised training institutes.

Subsidiaries

Indian Railways manufactures much of its rolling stock and heavy engineering components at its six manufacturing plants, called Production Units, which are managed directly by the ministry. As with most developing economies, the main reason for this was the policy of import substitution of expensive technology related products when the general state of the national engineering industry was immature. Each of these six production units is headed by a General Manager, who also reports directly to the Railway Board.

There exist independent organisations under the control of the Railway Board for electrification, modernisation and research and design, each of which is headed by a General Manager. A number of Public Sector Undertakings, which perform railway-related functions ranging from consultancy to ticketing, are also under the administrative control of the Ministry of railways.

Technical Details

Track

Indian railways uses four gauges, the 1,676mm broad gauge which is wider than the 1,435mm standard gauge; the 1,000mm metre gauge; and two narrow gauges, 762 mm (2 ft 6 in) and 610 mm (2 ft) . Track sections are rated for speeds ranging from 75 to 160 km/h. The total length of track used by Indian Railways was about 111,600 km (69,300 mi) while the total route length of the network was 63,273 km (39,316 mi) on 31 March 2008. About 28% of the route-kilometre and 42% of the total track kilometre was electrified on 31 March 2008.

Broad gauge is the predominant gauge used by Indian Railways. Indian broad gauge—1,676 mm (5 ft 6 in)—is the most widely used gauge in India with 96,851 km of track length (86.8% of entire track length of all the gauges) and 51,082 km of route-kilometre (80.7% of entire route-kilometre of all the gauges) on 31 March 2008.

In some regions with less traffic, the metre gauge (1,000mm) is common, although the Unigauge project is in progress to convert all tracks to broad gauge. The metre gauge had 11,676 km of track length (10.5% of entire track length of all the gauges) and 9,442 km of route-kilometre (14.9% of entire route-kilometre of all the gauges) on 31

March 2008. The Narrow gauges are present on a few routes, lying in hilly terrains and in some erstwhile private railways (on cost considerations), which are usually difficult to convert to broad gauge. Narrow gauges had a total of 2,749 route-kilometre on 31 March 2008. The Kalka-Shimla Railway, the Nilgiri Mountain Railway and the Darjeeling Himalayan Railway are three notable hill lines that use narrow gauge. Those three will not be converted under the Unigauge project.

The share of broad gauge in the total route-kilometre has been steadily rising, increasing from 47% (25,258 route-km) in 1951 to more than 83% in 2010 whereas the share of metre gauge has declined from 45% (24,185 route-km) to less than 13% in the same period and the share of narrow gauges has decreased from 8% to 3%. However, the total route-kilometre has increased by only 18% (by just 10,000 km from 53,596 route-km in 1951) in the last 60 years. This compares very poorly with Chinese railways, which increased from about 27,000 route-km at the end of second world war to about 90,000 route-km in 2010, an increase of more than three-fold. More than 28,000 route-km (34% of the total route-km) of Chinese railway is electrified compared to only about 18,000 route-km of Indian railways. This is an indication of the poor state of Indian railways where the funds allocated to new railway lines are meagre, construction of new uneconomic railway lines are taken up due to political interference without ensuring availability of funds and the projects incur huge cost and time overruns due to poor project-management and paucity of funds.

Sleepers (ties) used are made of prestressed concrete, or steel or cast iron posts, though teak sleepers are still in use on few older lines. The prestressed concrete sleeper is in wide use today. Metal sleepers were extensively used before the advent of concrete sleepers. Indian Railways divides the country into four zones on the basis of the range of track temperature. The greatest temperature variations occur in Rajasthan, where the difference may exceed 70°C.

Traction

As of March 2008, 18,274 km of the total 63,273 km route length is electrified. Since 1960, almost all electrified sections on IR use 25,000 V AC traction through overhead catenary delivery. A major exception is the entire Mumbai section, which uses 1,500 V DC. and is currently undergoing change to the 25,000 V AC system. Another exception is the Kolkata Metro, which uses 750 V DC delivered through a third rail. Traction voltages are changed at two places close to Mumbai. Central Railway trains passing through Igatpuri switch from AC to DC using a neutral section that may be switched to either voltage while the locomotives are decoupled and swapped. Western Railway trains switch power on the fly, in a section between Virar (DC) and Vaitarna (AC), where the train continues with its own momentum for about 30 m through an unelectrified section of catenary called a *dead zone.* All electric engines and EMUs operating in this section are the necessary AC/DC dual system type (classified "WCAM" by Indian Railways).

Services

Passenger

Indian Railways operates about 9,000 passenger trains and transports 20 million passengers daily across twenty-eight states and two union territories. Sikkim, Arunachal Pradesh, and Meghalaya are the only states not connected by rail. A standard passenger train consists of eighteen coaches, but popular trains can have up to 26 coaches. Coaches are designed to accommodate anywhere from 18 to 108 passengers, but during the holiday seasons and/or on busy routes, more passengers may travel in unreserved coaches. Most regular trains have coaches connected through vestibules. However, 'unreserved coaches' are not connected with the rest of the train via any vestibule.

Further, other AC classes can have 2 or 3 tier berths, with higher prices for the former, 3-tier non-AC coaches or 2nd class seating coaches, which are popular among passengers going on shorter journeys.

In air-conditioned sleeper classes passengers are provided with sheets, pillows and blankets. Meals and refreshments are provided, to all the passengers of reserved classes, either through the on-board pantry service or through special catering arrangements in trains without pantry car. Unreserved coach passengers have options of purchasing from licensed vendors either on board or on the platform of intermediate stops.

The amenities depend on the popularity and length of the route. Lavatories are communal and feature both the Indian style as well as the Western style.

The following table lists the classes in operation. Not all classes may be attached to a rake though.

Class	*Description*
1A	The First class AC: This is the most expensive class, where the fares are on par with airlines. Bedding is included with the fare in IR. This air conditioned coach is present only on popular routes between metropolitan cities and can carry 18 passengers. The coaches are carpeted, have sleeping accommodation and have privacy features like personal coupes.
2A	AC-Two tier: Air conditioned coaches with sleeping berths, ample leg room, curtains and individual reading lamps. Berths are usually arranged in two tiers in bays of six, four across the width of the coach then the gangway then two berths longways, with curtains provided to give some privacy from those walking up and down. Bedding is included with the fare. A broad gauge coach can carry 48 passengers.
FC	First class: Same as 1AC, without the air conditioning. This class is not very common.
3A	AC three tier: Air conditioned coaches with sleeping berths. Berths are usually arranged as in 2AC but with three tiers across the width and two longways as before giving eight bays of eight. They are slightly less well appointed, usually no reading lights or curtained off gangways. Bedding is included with fare. It carries 64 passengers in broad gauge.
3E	AC three tier (Economy): Air conditioned coaches with sleeping berths, present in Garib Rath Trains. Berths are usually arranged as in 3AC but with three tiers across the width and three longways. They are slightly less well appointed, usually no reading lights or curtained off gangways. Bedding is not included with fare.
CC	AC chair car: An air-conditioned seater coach with a total of five seats in a row used for day travel between cities.
EC	Executive class chair car: An air-conditioned seater coach with a total of four seats in a row used for day travel between cities.
SL	Sleeper class: The sleeper class is the most common coach, and usually ten or more coaches could be attached. These are regular sleeping coaches with three berths vertically stacked. In broad gauge, it carries 72 passengers per coach. Railways have modified certain Sleeper Coaches on popular trains to accommodate 81 passengers in place of regular 72 passengers. This was done in order to facilitate benefits like clear the Passenger rush and simultaneously earn more revenue. But this has got lukewarm response with criticism from the travellers and railways has decided to remove them.

2S	Seater class: same as AC Chair car, but with bench style seats and without the air-conditioning.
UR	Unreserved: The cheapest accommodation, with seats made of pressed wood and are rarely cushioned. Although entry into the compartment is guaranteed, a sitting seat is not guaranteed. Tickets issued are valid on any train on the same route if boarded within 24 hours of buying the ticket. These coaches are usually very crowded.
G	GaribRath: The cheapest air conditioned accommodation, the seats are same as the 3A coaches, but there are three berths at the side, unlike the other ac coaches, these coaches are very crowded.

At the rear of the train is a special compartment known as the guard's cabin. It is fitted with a transceiver and is where the guard usually gives the all clear signal before the train departs. A standard passenger rake generally has four general compartments, two at the front and two behind, of which one is exclusively for ladies. The exact number varies according to the demand and the route. A luggage compartment can also exist at the front or the back. In some trains a separate mail compartment is present. In long-distance trains a pantry car is usually included in the centre. A new class; Economy AC three tier is introduced in the Sealdah-New Delhi Duronto train.

Notable Trains and Achievements

There are two UNESCO World Heritage Sites on IR — the Chatrapati Shivaji Terminus and the Mountain railways of India. The latter is not contiguous, but actually consists of three separate railway lines located in different parts of India:

- The Darjeeling Himalayan Railway, a narrow gauge railway in West Bengal.
- The Nilgiri Mountain Railway, a metre gauge railway in the Nilgiri Hills in Tamil Nadu.
- The Kalka-Shimla Railway, a narrow gauge railway in the Shivalik mountains in Himachal Pradesh.

The *Palace on Wheels* is a specially designed train, frequently hauled by a steam locomotive, for promoting tourism in Rajasthan. On the same lines, the Maharashtra government introduced the *Deccan Odyssey* covering various tourist destinations in Maharashtra and Goa, and was followed by the Government of Karnataka which introduced the *Golden Chariot* train connecting popular tourist destinations in Karnataka and Goa. However, neither of them has been able to enjoy the popular success of the Palace on Wheels.

The *Samjhauta Express* is a train that runs between India and Pakistan. However, hostilities between the two nations in 2001 saw the line being closed. It was reopened when the hostilities subsided in 2004. Another train connecting Khokhrapar (Pakistan) and Munabao (India) is the *Thar Express* that restarted operations on February 18, 2006; it was earlier closed down after the 1965 Indo-Pak war. The Kalka Shimla Railway till recently featured in the Guinness Book of World Records for offering the steepest rise in altitude in the space of 96 kilometre.

The *Lifeline Express* is a special train popularly known as the "Hospital-on-Wheels" which provides healthcare to the rural areas. This train has a carriage that serves as an operating room, a second one which serves as a storeroom and an additional two that serve as a patient ward. The train travels around the country, staying at a location for about two months before moving elsewhere. Among the famous locomotives, the *Fairy Queen* is the oldest operating locomotive in the world today, though it is operated only for specials between Delhi and Alwar. *John Bull*, a locomotive older than Fairy Queen, operated in 1981 commemorating its 150 anniversary.

Kharagpur railway station also has the distinction of being the world's longest railway platform at 1,072 m (3,517 ft). The Ghum station along the Darjeeling Toy Train route is the second highest railway station in the world to be reached by a steam locomotive. The Mumbai–Pune Deccan Queen has the oldest running dining car in IR.

The *Himsagar Express*, between Kanyakumari and Jammu Tawi, has the longest run in terms of distance and time on Indian Railways network. It covers 3,745 km (2,327 mi) in about 74 hours and 55 minutes. The *Bhopal Shatabdi Express* is the fastest train in India today having a maximum speed of 150 km/h (93 mph) on the Faridabad–Agra section. The fastest speed attained by any train is 184 km/h (114 mph) in 2000 during test runs.

The *Rajdhani Express* and *Shatabdi Express* are the superfast, fully air-conditioned trains that give the unique opportunity of experiencing Indian Railways at its best. In July 2009, a new non-stop train service called Duronto Express was announced by the railway minister Mamata Banerjee.

Fares and Ticketing

Fares on the Indian Railways across categories are among the cheapest in the world. In the past few years, despite a recessionary environment, the Indian Railways have not raised fares on any class of service. On the contrary, there has been a minor dip in fares in some categories.

Ticketing services are available at all major and minor railway stations across India. In 2003, Indian Railways launched online ticketing services through the IRCTC website. Apart from E-tickets, passengers can also book I-tickets that are basically regular printed tickets, except that they are booked online and delivered by post. According to comScore, the Indian Railways website was the top visited Indian travel site in April 2010, with 7.7 million visitors.

Tourism

IRCTC takes care of the tourism operations of the Indian Railways. The Indian Railways operates several luxury trains such as Palace on Wheels, Golden Chariot, Royal Orient Express and Deccan Odyssey; that cater mostly to foreign tourists. For domestic tourists too, there are several packages available that cover various important tourist and pilgrimage destinations across India.

Freight

IR carries a huge variety of goods ranging from mineral ores, fertilizers and petrochemicals, agricultural produce, iron & steel, multimodal traffic and others. Ports and major urban areas have their own dedicated freight lines and yards. Many important freight stops have dedicated platforms and independent lines.

Indian Railways makes 70% of its revenues and most of its profits from the freight sector, and uses these profits to cross-subsidise the loss-making passenger sector. However, competition from trucks which offer cheaper rates has seen a decrease in freight traffic in recent years. Since the 1990s, Indian Railways has switched from small consignments to larger container movement which has helped speed up its operations. Most of its freight earnings come from such rakes carrying bulk goods such as coal, cement, food grains and iron ore. Indian Railways also transports vehicles over long distances. Trucks that carry goods to a particular location are hauled back by trains saving the trucking company on unnecessary fuel expenses.

Refrigerated vans are also available in many areas. The "Green Van" is a special type used to transport fresh food and vegetables. Recently Indian Railways introduced the special 'Container Rajdhani' or CONRAJ, for high priority freight. The highest speed notched up for a freight train is 100 kilometres per hour (62 mph) for a 4,700 metric tonne load. Recent changes have sought to boost the earnings from freight. A privatization scheme was introduced recently to improve the performance of freight trains. Companies are being allowed to run their own container trains.

The first length of an 11,000-kilometre (6,800 mi) freight corridor linking India's biggest cities has recently been approved. The railways has increased load limits for the system's 225,000 freight wagons by 11%, legalizing something that was already happening. Due to increase in manufacturing transport in India that was augmented by the increase in fuel cost, transportation by rail became advantageous financially. New measures such as speeding up the turnaround times have added some 24% to freight revenues.

Dedicated Freight Corridor

Ministry of Railways have planned to construct a new Dedicated Freight Corridor (DFC) covering about 2762 route km on two corridors, Eastern Corridor from Ludhiana to Sone Nagar and Western Corridor from Jawahar Lal Nehru Port Mumbai to Tughlakabad/Dadri along with interlinking of two corridors at Dadri. Upgrading of transportation technology, increase in productivity and reduction in unit transportation cost are the focus areas for the project.

"Dedicated Freight Corridor Corporation of India Limited (DFCC)" is a special purpose vehicle created to undertake planning & development, mobilization of financial resources and construction, maintenance and operation of the Dedicated Freight Corridors. DFCC has been registered as a company under the Companies Act 1956 on 30 October 2006.

Rail Budget and Finances

The Railway Budget deals with planned infrastructure expenditure on the railways as well as with the operating revenue and expenditure for the upcoming fiscal years, the public elements of which are usually the induction and improvement of existing trains and routes, planned investment in new and existing infrastructure elements, and the tariff for freight and passenger travel. The Parliament discusses the policies and allocations proposed in the budget. The budget needs to be passed by a simple majority in the Lok Sabha (Lower House).

The comments of the Rajya Sabha (Upper House) are non-binding. Indian Railways is subject to the same audit control as other government revenue and expenditures. Based on anticipated traffic and the projected tariff, requirement of resources for capital and revenue expenditure of railways is worked out. While the revenue expenditure is met entirely by railways itself, the shortfall in the capital (plan) expenditure is met partly from borrowings (raised by Indian Railway Finance Corporation) and the rest from Budgetary support from the Central Government. Indian Railways pays dividend to the Central Government for the capital invested by the Central Government.

As per the Separation Convention (on the recommendations of the Acworth Committee), 1924, the Railway Budget is presented to the Parliament by the Union Railway Minister, two days prior to the General Budget, usually around 26 February. Though the Railway Budget is separately presented to the Parliament, the figures relating to the receipt and expenditure of the Railways are also shown in the General Budget, since they are a part and parcel of the total receipts and expenditure of the Government of India. This document serves as a balance sheet of operations of the Railways during the previous year and lists out plans for expansion for the current year.

The formation of policy and overall control of the railways is vested in Railway Board, comprising the Chairman, the Financial Commissioner and other functional members of Traffic, Engineering, Mechanical, Electrical and Staff departments. Indian Railways, which a few years ago was operating at a loss, has, in recent years, been generating positive cash flows and been meeting its dividend obligations to the government, with (unaudited) operating profits going up substantially. The railway reported a cash surplus of INR 9000 cr in 2005, INR 14000 cr in 2006, INR 20,000 cr in 2007 and INR 25,000 cr for the 2007-2008 fiscal year. Its operating ratio improved to 76% while, in the last four years, its plan size increased from INR 13,000 cr to INR 30,000 cr.

The proposed investment for the 2008-2009 fiscal year is INR 37,500 cr, 21% more than for the previous fiscal year. Budget Estimates-2008 for Freight, Passenger, Sundry other Earnings and other Coaching Earnings have been kept at INR 52,700 cr, INR 21,681 cr, INR 5,000 cr and INR 2,420 cr respectively. Maintaining an overall double digit growth, Gross Traffic Earnings have been projected as INR 93,159 crore in 2009-10 (19.1 billion USD at current rate), exceeding the revised estimates for the current fiscal by INR 10,766 crore. Around 20% of the passenger revenue is earned from the upper class segments of the passenger segment (the air-conditioned classes).

The Sixth Pay Commission was constituted by the Government of India in 2005 to review the pay structure of government employees, and submitted its recommendations in April 2008. Based on its recommendations, the salaries of all Railways officers and staff were to be revised with retrospective effect w.e.f. January 1, 2006, resulting in an expenditure of over Rs. 13000 crore in 2008-09 and Rs. 14000 crore in 2009-10. Consequently, staff costs have risen from 44% of ordinary working expenses to 52%.

Issues

Many railway stations are in gross disrepair, dirty, outdated and overcrowded, especially when compared to stations in developed countries. Sometimes passengers are seen on trains hanging out windows and even on the roof creating safety problems. The interior of many train compartments are poorly maintained from rust, dirt and common wear and tear. Given the political infighting, corruption and inefficiencies, it is understandable that there are overcrowding, cleanliness and other maintenance issues. Although accidents such as derailment and collisions are less common in recent times, many are run over by trains, especially in crowded areas. Indian Railways have accepted the fact that given the size of operations, eliminating accidents is an unrealistic goal, and at best they can only minimize the accident rate. Human error is the primary cause, leading to 83% of all train accidents in India.

While accident rates are low-0.55 accidents per million train kilometre, the absolute number of people killed is high because of the large number of people making use of the network. While strengthening and modernisation of railway infrastructure is in progress, much of the network still uses old signalling and has antiquated bridges. Lack of funds is a major constraint for speedy modernisation of the network, which is further hampered by diversion of funds meant for infrastructure to lower-prioritised purposes due to political compulsions. In order to solve this problem, the Ministry of Railways in 2001 created a non-lapsible safety fund of Rs. 17000 crore exclusively for the renewal of overaged tracks, bridges, rolling stock and signalling gear. In 2003, the Ministry also prepared a Corporate Safety Plan for the next ten years with the objective of realising a vision of an accident-free and casualty-free railway system. The plan, with an outlay of Rs. 31835 crore, also envisaged development of appropriate technology for higher level of safety in train operation.

Reforms and Upgrades

Outdated communication, safety and signalling equipment, which used to contribute to failures in the system, is being updated with the latest technology. A number of train accidents happened on account of a system of manual signals between stations, so automated signalling is getting a boost at considerable expense. It is felt that this would be required given the gradual increase in train speeds and lengths, that would tend to make accidents more dangerous. In the latest instances of signalling control by means of interlinked stations, failure-detection circuits are provided for each track circuit and signal circuit with notification to the signal control centres in case of problems. Though currently available only in a small subset of the overall IR system, anti-collision devices are to be extended to the entire system. Aging colonial-era bridges and century-old tracks also require regular maintenance and upgrading.

The fastest trains of Indian Railways, Rajdhani Express and Shatabadi Express face competition from low-cost airlines since they run at a maximum speed of only 150 kilometres per hour (93 mph). At least six corridors are under consideration for the introduction of high speed bullet trains to India with expert assistance from France and Japan.

IR is in the process of upgrading stations, coaches, tracks, services, safety, and security, and streamlining its various software management systems including crew scheduling, freight, and passenger ticketing. Crew members will be able to log in using biometric scanners at kiosks while passengers can avail themselves of online booking. Initially, various upgrade and overhaul work will be performed at more than five hundred stations, some of it by private contract. All metre gauge lines in the country will be converted to broad gauge.

New LHB stainless steel coaches, manufactured in India, have been installed in Rajdhani and Shatabdi express trains. These coaches enhance the safety and riding comfort of passengers besides having more carrying capacity, and in time will replace thousands of old model coaches throughout Indian Railways.

More durable and conforming polyurethane paint is now being used to enhance the quality of rakes and significantly reduce the cost of repainting. Improved ventilation and illumination are part of the new scheme of things, along with the decision to install air brake systems on all coaches. New manufacturing units are being set up to produce state-of-the-art locomotives and coaches. IR is also expanding its telemedicine network facilities to further give its employees in far-flung and remote areas access to specialized medicine. IR has also piloted Internet connectivity on the Mumbai-Ahmedabad Shatabdi

Express, powered by Techno Sat Communications It is estimated that modernisation of IR and bringing it up to international standards would require US$280 billion in new upgrades and investment from 2010 to 2020.

Sanitation in trains and stations throughout the system is getting more attention with the introduction of eco-friendly, discharge-free, green (or bio-) toilets developed by IIT Kanpur. Updated eco-friendly refrigerant is being used in AC systems while fire detection systems will be installed on trains in a phased manner. New rodent-control and cleanliness procedures are working their way into the many zones of IR. Central Railway's 'Operation Saturday' is gradually making progress, station by station, in the cleanup of its Mumbai division.

Augmentation of capacity has also been carried out in order to meet increasing demand. The number of coaches on each train have been increased to 24, from 16, which increased costs by 28% but increased revenues by 78%. The railways were permitted to carry 68 tons per wagon, up from the earlier limit of 54 tons per wagon, thereby cutting costs.

The turnaround time for freight wagons was reduced from 7 days to 5 by operating the goods shed 24X7, electrifying every feeder line (this reduced time spent switching the engine from diesel to electric or from electric to diesel). Reducing the turnaround time meant that the Railways could now load 800 trains daily, instead of 550 trains daily. The minimum tonnage requirements were reduced allowing companies to unload their cargo at multiple stops.

Mumbai Suburban Railway

The Mumbai Suburban Railway system, part of the public transport system of Mumbai, is provided for by the state-run Indian Railways' two zonal Western Railways and Central Railways. The system carries more than 6.9 million commuters on a daily basis and constitutes more than half of the total daily passenger capacity of the Indian Railways itself. It has one of the highest passenger densities of any urban railway system in the world. The trains plying on its routes are commonly referred to as local trains or simply as locals by the general populace.

The Mumbai Suburban Railway, as well as Indian Railways, are an offshoot of the first railway to be built by the British in India in April 1853, and was also the oldest railway system in Asia. The first train ran between Mumbai and Thane, a distance of 34 km. The

Bombay Railway History Group has been striving to document railway heritage along this line.

Due to the geographical spread of the population and location of business areas, the rail network is the principal mode of mass transport in Mumbai. As Mumbai's population swelled from a heavy inflow of migrants in recent decades, frequent overcrowding has become a serious issue, and numerous safety concerns have been raised over the years. A metro system and a monorail system are under construction in Mumbai to ease the travelling conditions in the Suburban network.

Facilities

Spread over 464 route kilometres, The Suburban Railway system operates on 1500 V DC/25000 V AC (Virar-Borivali & Kasara-Titwala) power supply from overhead catenary lines. The suburban services are run by electric multiple units (EMUs). 191 rakes (train sets) of 9-car & 12-car composition are utilised to run 2342 train services, carrying 6.94 million passengers per day.

Zones and Corridors

Two zonal Railways, the Western Railway (WR) and the Central Railway (CR), operate the Mumbai Suburban Railway system. At present, the fast corridors on Central Railway as well as Western Railway are shared for long distance (main line) and freight trains.

Western Line

Two corridors (one local and the other through) on Western Railway run northwards from Churchgate terminus parallel to the west coast up to Dahanu Road (120 km). These corridors are popularly referred to as 'Western Line' by the locals mainly because it is operated and owned by the Western Railways. Electric Multiple Units (EMUs) ply between Churchgate and Virar (64 km), while Mainline Electrical Multiple Units (MEMUs) service the section beyond Virar till Dahanu Road (60 km). MEMUs also operate between Dahanu Road and Panvel via a branch line from Vasai Road. There are EMU carsheds at Mumbai Central and Kandivali. A repair shop for EMUs is situated at Mahalaxmi. Western railway's EMU fleet consists of EMUs running on DC (1.5 kV) power as well as those running as dual-current i.e. AC (25 kV) and DC (1.5 KV) power. EMUs are 9 car, 12 car or 15 car formations and are differentiated as slow and fast locals. Slow trains halt at all stations, while fast ones halt at important stations only and are preferable over longer distances. Trains usually start from and terminate at important stations.

Safety Issues

Overcrowding

Due to its extensive reach across the Mumbai Metropolitan Region, and its intensive use by the local urban population, the Mumbai Suburban Railway suffers from some of the most severe overcrowding in the world. Over 4,500 passengers are packed into a 9-car rake during peak hours, as against the rated carrying capacity of 1,700. This has resulted in what is known as *Super-Dense Crush Load* of 14 to 16 standing passengers per square meter of floor space. Trains on the suburban line are on average more than 4 minutes apart, contributing to the problem of overcrowding. The impending introduction of new higher speed rakes may help address the issue.

Tourists

It has been advised for safety concerns for tourists to avoid the trains during weekdays, or at least during the morning and evening peak hours. Avoid travelling from north to south between 8 am and 11 am in the morning and from south to north between 6 pm and 9 pm at night. The best way to enjoy the trains is on Sundays when they are relatively empty. However, watch out for Sundays when work is done on the tracks, as it could mean that trains are still crowded on a Sunday. During the work day, beware of getting on the express trains or 'fast trains' as they are called denoted on stations by 'F', especially the trains to Virar.

Fatalities

More than 3,500 people die on the Mumbai suburban railway tracks annually due to unsafe riding on trains or trespassing on railway tracks or as a result of suicide attempts. This is believed to be the highest number of fatalities per year on any urban or suburban railway system. Most of the deaths are of passengers crossing the tracks on foot, instead of using the footbridges provided for going from one platform to another, and are hit by passing trains. Some passengers die when they sit on train roofs to avoid the crowds and are electrocuted by the overhead electric wires, or hang from doors and window bars. To reduce the risk of such fatalities, automatic doors will be installed on all rakes by 2016 along with longer platforms and more frequent trains.

Central and Western Railway was forced to release under the Right to Information Act that at least 20,706 people have died in the last five years; an average of 10 each day. The request was filed by Mumbai activist Chetan Kothari.

According to The Times of UK, Mumbai's local railway network was one of the deadliest in the world: a record 17 people died every weekday on the city's suburban railway network in 2008. Most deaths were people being run over while trespassing on the tracks. The next biggest cause of death was of passengers who fell (or were pushed) from carriages and are often dangerously full. Another 41 people perished after being bludgeoned by trackside poles while hanging out of overcrowded trains. Twenty-one were electrocuted by power cables when they sat on the roof.

International

Rail links between India and neighbouring countries are not well-developed. Two trains operate to Pakistan-the *Samjhauta Express* between Delhi and Lahore, and the *Thar Express* between Jodhpur and Karachi. Bangladesh is connected by a bi-weekly train, the *Maitree Express*. Nominal rail links to Nepal exist — passenger services between Jaynagar and Bijalpura, and freight services between Raxaul and Birganj. No rail link exists at present with Myanmar, but a railway line is to be built through from Jiribam (in Manipur) to Tamu through Imphal and Moreh. The construction of this missing link, as per the feasibility study conducted by the Ministry of External Affairs through RITES Ltd, is estimated to cost Rs 2,941 crore. Rail links with Bhutan have been proposed. No rail link exists with either China or Sri Lanka, however in the disputed region of Gilgit-Baltistan, the current controller of this area, Pakistan has proposed a rail link with China across the Khunjerab Pass.

Road

India has a network of National Highways connecting all the major cities and state capitals, forming the economic backbone of the country. As of 2005, India has a total of 66,590 km (41,377 mi) of National Highways, of which 200 km (124 mi) are classified as expressways. Under National Highways Development Project (NHDP), work is under progress to equip some of the important national highways with four lanes; also there is a plan to convert some stretches of these roads to six lanes. However congestion and bureaucratic delays enroute ensure that trucking goods from Gurgaon to the port in Mumbai can take up to 10 days.

As per the National Highways Authority of India, about 65% of freight and 80% passenger traffic is carried by the roads. The National Highways carry about 40% of total road traffic, though only about 2%

of the road network is covered by these roads. Average growth of the number of vehicles has been around 10.16% per annum over recent years. Highways have facilitated development along the route and many towns have sprung up along major highways.

All national highways are metalled, but very few are constructed of concrete, the most notable being the Mumbai-Pune Expressway. In recent years construction has commenced on a nationwide system of multi-lane highways, including the Golden Quadrilateral and North-South and East-West Corridors which link the largest cities in India. In 2000, around 40% of villages in India lacked access to all-weather roads and remained isolated during the monsoon season. To improve rural connectivity, *Pradhan Mantri Gram Sadak Yojana* (Prime Minister's Rural Road Program), a project funded by the Central Government with the help of World Bank, was launched in 2000 to build all-weather roads to connect all habitations with a population of 500 or above (250 or above for hilly areas).

As per 2009 estimates, the total road length in India is 3,320,410 km (2,063,210 mi); making the Indian road network the third largest road network in the world. At 0.66 km of highway per square kilometre of land the density of India's highway network is higher than that of the United States (0.65) and far higher than that of China's (0.16) or Brazil's (0.20).

Buses are an important means of public transport in India, particularly in the countryside and remote areas where the rail network cannot be accessed and airline operations are few or non-existent. Due to this social significance, public bus transport is predominantly owned and operated by public agencies, and most state governments operate bus services through a State Road Transport Corporation. These corporations, introduced in the 1960s and 1970s, have proven extremely useful in connecting villages and towns across the country.

Water and Sea Transport

Maritime transportation in India is managed by the Shipping Corporation of India, a government-owned company that also manages offshore and other marine transport infrastructure in the country. It owns and operates about 35% of Indian tonnage and operates in practically all areas of shipping business servicing both national and international trades. It has a fleet of 79 ships of 27 lakh GT (48 lakh DWT) and also manages 53 research, survey and support vessels of 1.2 Lakh GT (0.6 Lakh DWT) on behalf of various government departments and other organisations. Personnel are trained at the

Maritime Training Institute in Mumbai, a branch of the World Maritime University, which was set up in 1987. The Corporation also operates in Malta and Iran through joint ventures.

Ports

The ports are the main centres of trade. In India about 95% of the foreign trade by quantity and 70% by value takes place through the ports. There are twelve major ports: Kolkata (including Haldia), Paradip, Vishakapatnam, Ennore, Chennai, Tuticorin, Kochi, New Mangalore, Mormugao, Navi Mumbai, Mumbai and Kandla. Other than these, there are 187 minor and intermediate ports, 43 of which handle cargo.

The distinction between major and minor ports is not based on the amount of cargo handled. The major ports are managed by port trusts which are regulated by the central government. They come under the purview of the Major Port Trusts Act, 1963. The minor ports are regulated by the respective state governments and many of these ports are private ports or captive ports. The total amount of traffic handled at the major ports in 2005-2006 was 382.33 Mt.

Waterways

India has an extensive network of inland waterways in the form of rivers, canals, backwaters and creeks. The total navigable length is 14,500 kilometres (9,000 mi), out of which about 5,200 km (3,231 mi) of river and 485 km (301 mi) of canals can be used by mechanised crafts. Freight transport by waterways is highly underutilised in India compared to other large countries.

The total cargo moved by inland waterways is just 0.15% of the total inland traffic in India, compared to the corresponding figures of 20% for Germany and 32% for Bangladesh. Cargo transport in an organised manner is confined to a few waterways in Goa, West Bengal, Assam and Kerala. The Inland Waterways Authority of India (IWAI) is the statutory authority in charge of the waterways in India. It does the function of building the necessary infrastructure in these waterways, surveying the economic feasibility of new projects and also administration and regulation. The following waterways have been declared as National Waterways:

- National Waterway 1-Allahabad-Haldia stretch of the Ganga-Bhagirathi-Hooghly river system with a total length of 1,620 kilometres (1,010 mi) in October 1986.

- National Waterway 2-*Saidiya*-Dhubri stretch of the Brahmaputra river system with a total length of 891 kilometres (554 mi) in 1988.
- National Waterway 3-Kollam-Kottapuram stretch of the West Coast Canal along with Champakara and Udyogmandal canals, with a total length of 205 kilometres (127 mi) in 1993.
- National Waterway 4-Bhadrachalam-Rajahmundry and Wazirabad-Vijaywada stretch of the Krishna-Godavari river system along with the Kakinada-Puducherry canal network, with a total length of 1,095 km (680 mi) in 2007.
- National Waterway 5-*Mangalgadi*-Paradeep and Talcher-*Dhamara* stretch of the Mahanadi-Brahmani river system along with the East Coast Canal, with a total length of 623 km (387 mi) in 2007.

Aviation

Rapid economic growth in India has made air travel more affordable. Air India, India's flag carrier, presently operates a fleet of 159 aircraft and plays a major role in connecting India with the rest of the world. Several other foreign airlines connect Indian cities with other major cities across the globe.

Kingfisher Airlines, Air India and Jet Airways are the most popular brands in domestic air travel in order of their market share. These airlines connect more than 80 cities across India and also operate overseas routes after the liberalisation of Indian aviation. However, a large section of country's air transport system remains untapped, even though the Mumbai-Delhi air corridor was ranked 6th by the Official Airline Guide in 2007 among the world's busiest routes.

India's vast unutilised air transport network has attracted several investments in the Indian air industry in the past few years. More than half a dozen low-cost carriers entered the Indian market in 2004-05. Major new entrants include Air Deccan, Kingfisher Airlines, SpiceJet, GoAir, Paramount Airways and IndiGo Airlines. To meet India's rapidly increasing demand for air travel, Air India recently placed orders for more than 68 jets from Boeing for 7.5 billion USD while Indian placed orders for 43 jets from Airbus for 2.5 billion USD. Jet Airways, India's largest private carrier, has invested millions of dollars to increase its fleet, but this has been put on hold due to the recent economic slowdown. This trend is not restricted to traditional air carriers in India. IndiGo Airlines entered the limelight when it announced orders for 100 Airbus A320s worth 6 billion USD during the Paris Air Show; the highest by any Asian domestic carrier. Kingfisher Airlines became the first Indian air carrier in June 15, 2005 to order Airbus A380 aircraft. The total deal with Airbus was worth 3 billion USD.

Airports

There are more than 335 (2008 est.) civilian airports in India-250 with paved runways and 96 with unpaved runways and more than 20 international airports in the Republic of India. The Indira Gandhi International Airport and the Chhatrapati Shivaji International Airport handle more than half of the air traffic in South Asia.

Heliports

As of 2007, there are 30 heliports in India. India also has the world's highest helipad at the Siachen Glacier a height of 6400 metre (21,000 ft) above mean sea level.

Pawan Hans Helicopters Limited is a public sector company that provides helicopter services to ONGC to its off-shore locations, and also to various State Governments in India, particularly in North-east India.

Pipelines:

- Length of pipelines for crude oil is 20,000 km (12,427 mi).
- Length of Petroleum products pipeline is 268 km (167 mi).
- Length of Natural gas pipelines is 1,700 km (1,056 mi).

The above information was calculated in 2008.

Environmental Issues and Impact

The National capital New Delhi has one of the largest CNG based transport systems as a part of the drive to bring down pollution. In spite of these efforts it remains the largest contributor to the greenhouse gas emissions in the city. The CNG Bus manufacturers in India are Ashok Leyland, Tata Motors, Swaraj Mazda and Hindustan Motors.

The Karnataka State Road Transport Corporation was the first State Transport Undertaking in India to utilise bio-fuels and ethanol-blended fuels. KSRTC took an initiative to do research in alternative fuel forms by experimenting with various alternatives— blending diesel with biofuels such as honge, palm, sunflower, groundnut, coconut and sesame. In 2009, the corporation decided to promote the use of biofuel buses.

In 1998, the Supreme Court of India published a Directive that specified the date of April 2001 as deadline to replace or convert all buses, three-wheelers and taxis to Compressed Natural Gas.

Transport in Delhi

Delhi has significant reliance on its transport infrastructure. The city has developed a highly efficient public transport system with the introduction of the Delhi Metro, which is undergoing a rapid modernization and expansion.

There are 5.5 million registered vehicles in the city, which is the highest in the world among all cities most of which do not follow any pollution emission norm (within municipal limits), while the Delhi metropolitan region (NCR Delhi) has 11.2 million vehicles. Delhi and NCR lose nearly 42 crore (420 million) man-hours every month while commuting between home and office through public transport, due to the traffic congestion.. Therefore serious efforts, including a number of transport infrastructure projects, under way to encourage usage of public transport in the city.

Overview

Public transport in the metropolis includes the Delhi Metro, the Delhi Transport Corporation bus system, auto-rickshaws, cycle-rickshaws and taxis. With the introduction of Delhi Metro, a rail-based mass rapid transit system, rail-based transit systems have gained ground. Other means of transit include suburban railways, inter-state bus services and private taxis which can be rented for various purposes. However, buses continue to be the most popular means of transportation for intra-city travel, they cater to about 60% of the total commuting requirements.

Private vehicles account for 30% of the total demand for transportation, while the rest of the demand is met largely by auto-rickshaws, taxis, rapid transit system and railways.

Indira Gandhi International Airport (IGI) serves Delhi for both domestic and international air connections, and is situated in the south-western corner of the city. In 2005-2006, IGI recorded a traffic of more than 20.44 million passengers. (Both Domestic and International), Heavy air-traffic has stressed on the need for a secondary airport, which is expected to come-up in the form of Taj International Airport near Greater Noida, alongside Delhi-Agra highway. The Delhi government is planning to have 413 km of metro, 292 km of BRT, and 50 km each of monorail and light rail by 2020.

Intra-City Transport

Road Transport

Transportation in Delhi is largely dependent upon roads. Railways, including rapid transit systems like Delhi Metro. Roads in Delhi are maintained by Municipal Corporation of Delhi (MCD), New Delhi Municipal Committee (NDMC), Delhi Cantonment Board (DCB), Public Works Department (PWD) and Delhi Development Authority (DDA). At 1749 km of road length per 100 km^2, Delhi has one of the highest road densities in India. Major roadways include the Ring Road and the Outer Ring Road, which had a traffic density of 110,000 vehicles per day in 2001. Total road length of Delhi was 28,508 km including 388 km of National Highways.

Major road-based public transport facilities in Delhi are provided by DTC buses, auto-rickshaws, taxis and cycle-rickshaws.

Buses

Delhi has one of India's largest bus transport systems. Buses are the most popular means of transport catering to about 60% of Delhi's total demand. Buses are operated by the state-owned Delhi Transport Corporation (DTC), which owns largest fleet of Compressed Natural Gas (CNG)-fueled buses in the world, private Blueline bus operators and several chartered bus operators. It is mandatory for all private bus operators to acquire a permit from the State Transport Authority. The buses traverse various well-defined intra-city routes. Other than regular routes, buses also travel on Railway Special routes; Metro Feeder routes. Mudrika (Ring) and Bahri Mudrika (Outer Ring) routes along Ring and Outer-Ring road respectively are amongst the longest intra-city bus routes in the world.

With the introduction of Bus Rapid Transit (BRT) and the development of dedicated corridors for the service, bus service is set to improve. The DTC has started introducing air-conditioned buses and brand new low-floor buses (with floor height of 400 mm and even higher on one third area as against 230 mm available internationally.) on city streets to replace the conventional buses. A revamp plan is underway to improve bus-shelters in the city and to integrate GPS systems in DTC buses and bus stops so as to provide reliable information about bus arrivals.

In 2007, after public uproar concerning the large number of accidents caused by privately-owned Blueline buses, the Delhi government, under pressure from the Delhi High Court decided that all Blueline Buses shall be phased out and be eventually replaced by low floor buses of the state-owned DTC. The Delhi Government has decided to expedite this process and will procure 6,600 low floor buses for the DTC by commonwealth games next year.

By 2010, Delhi will have over 8000 buses, of which Delhi Transport Corporation will provide 6000 while 2000 would be blueline buses, 3125 will be low-floor,1100 semi low floor and 1000 of them would be air-conditioned. Few buses would have GPS to prevent them from straying to other routes. The city already has 655 low-floor AC and non-AC buses. The bus routes are also being increased to 670 from the current 357 routes. Delhi plans to add at least 2500 of these new buses by the end of 2009. The city has been divided into 17 clusters. Bus services in each of these clusters will be run by private operators. The first cluster is to be awarded by September 2008. The first cluster has 32 routes, on which a total of 295 DTC and 270 private buses will run. The operators will be given the option of running 20 percent AC buses With the introduction of new buses, DTC will be recruiting 4000 drivers to run the new buses.

In November 2009, DTC piloted a program to introduce the smart card where the commuters would be able to pay the fare through the smart card. They have decided to install the machines in 10000 buses.

Auto-Rickshaws

The auto-rickshaws (popularly known as *Auto*) are an important and popular means of public transportation in Delhi, as they are cheaper than taxis. Hiring an *Auto* in Delhi is very tricky, as very few auto-drivers agree to standard meter charges. The typical method is to haggle for an agreeable rate. This rarely is a source of conflict, because the fares charged are modest and the ride ensures speedy arrival to the destination.

Taxis

Though easily available, taxis are not an integral part of Delhi public transport. The Indian Tourism Ministry and various private owners operate most taxis. The Tourism Ministry grants private companies permits to operate taxis.

Recently, Radio Taxis have started to gain ground in Delhi. Brands such as Mega Cabs, EasyCabs, etc., provide the on-call radio taxi service, which is slightly more expensive than conventional Black and Yellow taxis. Other than these two mentioned, companies such as Hertz Car Rental and Avis Car Rental provide rent-a-car service.

Cycle-Rickshaws

Cycle-Rickshaws are a popular mode of travel for short distance transits in the city. The pedal-powered rickshaws are easily available throughout the city and reckoned for being cheap and environment friendly. Often, tourists and citizens use them for joyrides, too. Of late, they have been phased out from the congested areas of Chandni Chowk because of their slow pace, which often leads to traffic snarls on the streets of Old Delhi. Still, they are the great source of public transport in Delhi.

Major Arteries

Inner Ring Road

Inner Ring Road is one of the most important "state highways" in Delhi. It is a 51 km long circular road, which connects important areas in Delhi. Owing to more than 2 dozen grade-separators/flyovers, the road is almost signal-free. The road is generally 8-laned with a few bottlenecks at certain stretches, which are being removed. The road has already achieved its carrying capacity of 110000 vehicles per day and would require an addition of more lanes to fulfil needs of increasing traffic by 2011.

Outer Ring Road

Outer Ring Road is another major artery in Delhi. The road which was almost neglected till early 2000s is now an important highway that links far-flung areas of Delhi. The road is 6-8 lane and has grade-separators and a large number are under construction as a part of project to make the artery signal free. The road along with the ring road forms a ring which intersects all the National Highways passing through Delhi.

Expressways and Highways

Delhi is connected by NH 1,2,8 and 24. It also has 3 expressways (6 and 8 lane) that connect it with its suburbs. 4 more expressways are also planned and are supposed to be finished by 2010.

Delhi-Gurgaon Expressway connects Delhi with one of its financial hubs, Gurgaon.

DND Flyway connects Delhi with its other financial hub, Noida. Noida is also to host the Indian grand Prix

Noida-Greater Noida Highway connects Noida with Greater Noida, which is an upcoming financial and commercial hub and is also to have a new international airport. The construction work for 135.6-km long Delhi Western Peripheral Expressway also known as the Kundli-Manesar-Palwal Expressway(KMP) is going on at full swing.[1]Kundli-Manesar-Palwal (KMP) expressway expected to become operational by June 2009, Delhi will be relieved of the congestion of heavy night traffic. It will act as a bypass for the night vehicles.

Ghazibad-Faridabad-Gurgaon Expressway is a bypass corridor for traffic coming from South West and going towards East. It is currently under construction. Faridabad Road is a 4 lane highway road which connects Faridabad, major suburb to Delhi. Upgradation to expressway is underway. Ghaziabad Road is a 4 lane highway road which connects Ghaziabad to Delhi. As the Commonwealth Village is located close by Yamuna bridge on this highway, underpasses and flyover being built will help facilitate traffic between the eastern areas of Delhi/Western UP and the rest of the city. If the underpass, flyovers and bridges are constructed in time they might be extended to Ghaziabad.

Rail Transport

Rail based transport in the city has started to gain-popularity with the introduction of Delhi Metro. Ring-Railway, which runs parallel to the Ring-Road system is another rail-based intra-city transport facility in Delhi.

Metro

Rapid increase of population coupled with large-scale immigration due to high economic growth has resulted in ever increasing demand for better transport, putting excessive pressure on the city's existent transport infrastructure. Like many other cities in the developing world, the city faces acute transport management problems leading to air pollution, congestion and resultant loss of productivity.

In order to meet the transportation demand in Delhi, the State and Union government started the construction of an ambitious Mass Rapid Transit system, known as [5] Delhi Metro in 1998.

The project started commercial operations on December 24, 2002. It has set many performance and efficiency standards ever since and is continuously expanding at a very rapid pace. As of 2010, the metro operates 5 lines with a total length of 111 km and 98 stations while several other lines are under construction.

The second phase (Phase II) which is under construction has 128 km of route length and is slated to be completed by 2010. Phases III (112 km) and IV (108.5 km) will be completed by 2015 and 2020 respectively, with the network totalling 413.8 km, making it longer than the London Underground. With further development of the city, the network will be further expanded by adding new lines, thus crossing 500 km by 2020.

Ring Railway

Ring railway is a circular rail network in Delhi, which runs parallel to the Ring Road and was conceived during the Asian Games of 1982. The system is not popular amongst people and a total failure as far as public transport is considered. The major reasons for failure of the system are lack of proper connectivity, less population density in areas of reach. The network is now utilized as a freight corridor and limited passenger train services are available during peak hours.

Inter-state Transport

Railway Connectivity

Delhi is connected to whole of the nation through Indian Railways vast network. New Delhi Railway Station which is one of the most busiest stations in Indian Railway system serves as headquarter of Northern Railways. A large load of inter-state transport is borne by railways. Major railway stations in the city include New Delhi Railway Station, Old Delhi Railway Station, Hazrat Nizamuddin Railway Station. A large number of local passenger trains connect Delhi to its sub-urban areas and thus provide convenient travel for daily commuters. Railways also share a large amount of freight traffic in Delhi.

Road

Highways: The city is believed to have highest road density in the country and is well connected to rest of the nation through five major national highways, namely NH 1, NH 2, NH 8, NH 10 and NH 24. The highways around city are being upgraded into expressways with ultra-modern facilities.

Bus services

Regular bus services are available from inter-state bus terminals in the city. The services are extended to all the northern states and the neighbouring areas of Delhi. Services are provided by state transport corporations and several private operators. The inter-state terminals in city are:

- Kashmiri Gate ISBT in Northern Delhi
- Anand Vihar ISBT in Trans-Yamuna area
- Sarai Kale Khan ISBT in South-East Delhi.

Airports

Indira Gandhi International Airport (IGI) serves Delhi for both domestic and international connections, and is situated in the southwestern corner of the city, alongside Delhi-Gurgaon Expressway. In the year 2006-2007, IGI recorded a traffic of 20.44 million passengers. It is currently the busiest airport in South Asia. It operates two terminals — Terminal 1 for domestic and Terminal 2 for international air travel.

The airport is witnessing massive expansion and modernisation by a consortium led by GMR. The airport will get a new integrated Terminal 3 by 2010. Terminals 4, 5 and 6 will be built in a phased manner. By 2024, airport will have four runways and will handle more than 100 million passengers per year, which is more than what Atlanta airport (world's busiest airport) handles now.

Apart from the expanded IGI airport, Delhi might also receive a second airport by 2012-2013. The airport, being named as Taj International Aviation Hub, is proposed be located in Jewar in Greater Noida. It would be 75 km from IGI airport.

Future Projects

There are many transport infrastructure projects underway in Delhi. Most have their deadlines set in late 2009 and early 2010, just before the 2010 Commonwealth Games. They are listed below-

Road:

- Bus Rapid Transit corridors
- Taxi systems
- Revamp of DTC bus fleet, to include semi-low-floor. low floor and air-conditioned buses.
- Development of Eastern and Western Peripheral Highways to take off the load of inter-state traffic from roads of Delhi.

Rail:

- Upgradation of New Delhi and Old Delhi railway stations of Northern Railways.
- Expansion of existing Delhi Metro network, including a super-fast Delhi Airport Express Line having maximum speed of 135 km/h line to connect to IGI Airport.
- Introduction of Monorail (45 km) and Light Rail Transit has been aborted till 2010

- Anand Vihar Railway Terminal (about to be completed) to reduce the train loads over Old Delhi Station and New Delhi Railway Station. Besides that the station will also serve the densely populated Eastern part of delhi, along with the neighbouring suburbs of Ghaziabad and Noida.

Air:

- Revamp of IGI Airport is underway to improve its infrastructure, passenger capacity and efficiency.
- A secondary airport is in planning stages and will come up near Greater Noida.

Delhi Traffic Police Transport Helpline

Owing to large amount of complaints from consumers, the Delhi Government in association with Delhi Traffic Police runs a manned transport helpline which can be reached at 011-23010101 while dialing from within the city. Citizens can make traffic related complaints and suggestions. One can also report traffic violations observed and misbehavior/refusal/overcharging by Autorickshaws, Buses and Taxis

National Highway Development and Planning in India

In India, the National Highways are the primary long-distance roadways. Most are maintained by the Government of India, others are operated under a public-private partnership by the private sector. Most are two-lane (one in each direction). They span about 67,000 km (42,000 mi), of which about 200 km (120 mi) are designated expressways and 10,000 km (6,200 mi) have four lanes or more. National highways constitute approximately 2% of the total road network of India, but carry nearly 40% of the total traffic. The National Highways Development Project, currently being implemented, seeks to massively expand India's highway network.

Historical National Highways

In ancient times the ruling monarchs constructed many brick roads in cities. The most famous highway of medieval India was the Grand Trunk Road. The Grand Trunk Road begins in Sonargaonnear Dhaka, Bangladesh and ends in Peshawar, Pakistan. It travels through or near many important cities of the subcontinent, including Dhaka in Bangladesh, Kolkata, Patna, Varanasi, Kanpur, Agra, Delhi, Panipat, Ludhiana, Jalandhar, and Amritsar in India, and Lahore and Peshawar in Pakistan. In the 19th century, the British upgraded the existing highway network, and built roads in treacherous areas such as the Western Ghats.

Current System

India has 67,000 km (42,000 mi) of highways connecting all the major cities and state capitals. Most are two-lane highways with paved roads. In developed areas they may broaden to four lanes, while near large cities, they may expand to eight lanes. In most developed states, the roads are free of potholes. In less-developed states and sparsely populated areas, inadequate maintenance and the harsh monsoon results in potholed roads. Very few of India's highways are built from concrete. As of 2010, 19,064 km (11,846 mi) of the National Highway system still consists of single-laned roads. The government is currently working to ensure that by December 2014 the entire National Highway network consists of roads with two or more lanes.

India has the distinction of having the world's second highest-altitude motor highwayLeh-Manali Highway, connecting Shimla to Leh in Ladakh, Kashmir.

Highways form the economic backbone of the country. Highways have often facilitated development along their routes, and many new towns have sprung up along major highways. Highways also have large numbers of small restaurants and inns (known as *dhabas*) along their length. They serve popular local cuisine and serve as truck stops.

Recent Developments

Under former Prime Minister Atal Behari Vajpayee, India launched a massive program of highway upgrades, called the National Highway Development Project (NHDP), in which the main north-south and east-west connecting corridors and highways connecting the four metropolitan cities have been fully paved and widened into 4-lane highways. Some of the busier National Highway sectors in India have been converted to four or six lane expressways – for example, Delhi-Agra, Delhi-Jaipur, Ahmedabad-Vadodara, Mumbai-Pune, Mumbai-Surat, Bangalore-Mysore, Bangalore-Chennai, Chennai-Tada, Delhi-Meerut Hyderabad-Vijayawada and Guntur-Vijayawada. Phase V of the National Highway Development Project is to convert all 6,000 km (3,700 mi) of the Golden Quadrilateral Highways to 6-lane highways/expressways by 2012.

The National Highways Bill, passed in 1995, provides for private investment in the building and maintenance of the highways. Recently, a number of new roads have been classified as "National Highways" in a move to provide national connectivity to remote places. Bypasses have also recently been constructed around larger towns and cities to provide uninterrupted passage for highway traffic. The varied climactic, demographic, traffic, and sometimes political situation, prevents these highways from having a uniform character.

They range from fully-paved, six-lane roads in some areas, to unpaved stretches in remote places. Many National Highway's are still being upgraded or are under construction. There are long National Highway's to connect the metros together, as well as short spurs off the highway to provide connectivity to nearby ports or harbors. The longest National Highway is the NH7, which runs between Varanasi in Uttar Pradesh to Kanyakumari in Tamil Nadu, at the southernmost point of the Indian mainland, covering a distance of 2,369 km (1,472 mi), and passes through Hyderabad and Bangalore. The shortest National Highway is the NH47A, which spans 6 km (3.7 mi), to the Ernakulam-Kochi Port.

Indian Road Network

Indian Road Network

Class	Length (km)
Access Controlled Expressways	200 km (120 mi)
4-6 lane Divided Highways	
(with service rd in crowded areas)	10,000 km (6,200 mi)
National Highways	66,590 km (41,380 mi)
State Highways	131,899 km (81,958 mi)
Major district roads	467,763 km (290,654 mi)
Rural & other roads	2,650,000 km (1,650,000 mi)
Total (approx)	3,300,000 km (2,050,000 mi)

National Highways Authority of India

The National Highways Authority of India (NHAI) is an autonomous agency of the Government of India, responsible for management of a network of over 60,000 km of National Highways in India. The Authority is a nodal agency of the Ministry of Road Transport and Highways.

Establishment

The NHAI was created through the promulagation of the National Highways Authority of India Act, 1988. *In February 1995, the Authority was formally made an autonomous body. It succeeded the erstwhile Ministry of Surface Transport.* It is responsible for the development, maintenance, management and operation of National Highways, totaling over 70,548 km in length.

Projects

The NHAI has the mandate to implement the National Highway Development Project (NHDP). The NHDP is under implementation in Phases.

- Phase I: Approved in December 2000, at an estimated cost of INR 300 Billion, it included the Golden Quadrilateral (GQ), portions of the NS-EW Corridors, and connecitivity of major ports to National Highways.
- Phase II: Approved in December 2003, at an estimated cost of INR 343 Billion, it included the completion of the NS-EW corridors and another 486 km of highways.
- Phase IIIA: This phase was approved in March 2005, at an estimated cost of INR 222 Billion, it includes an upgrade to 4-lanes of 4,035 km of National Highways.
- Phase IIIB: This was approved in April 2006, at an estimated cost of INR 543 Billion, it includes an upgrade to 4-lanes of 8,074 km of National Highways.
- Phase V: Approved in October 2006, it includes upgrades to 6-lanes for 6,500 km, of which 5,700 km is on the GQ. This phase is entirely on a DBFO basis.
- Phase VI: This phase, approved in November 2006, will develop 1,000 km of expressways at an estimated cost of INR 167 Billion.
- Phase VII: This phase, approved in December 2007, will develop ring-roads, bypasses and flyovers to avoid traffic bottlenecks on selected stretches at a cost of INR 167 Billion.

National Highways Development Project

The National Highways Development Project is a project to upgrade, rehabilitate and widen major highways in India to a higher standard. The project was implemented in 1998. "National Highways" account for only about 2% of the total length of roads, but carry about 40% of the total traffic across the length and breadth of the country.

This project is managed by the National Highways Authority of India under the Ministry of Road, Transport and Highways. The NHAI has implemented US$ 71 billion for this project, as of 2006.

Phases

The project is composed of the following phases:

- Phase I: The Golden Quadrilateral (GQ; 5,846 km) connecting the four major cities of Delhi, Mumbai, Chennai and Kolkata. This project connecting four metro cities, would be 5,846 km.

Total cost of the project is Rs300 billion (US$6.8 billion), funded largely by the government's special petroleum product tax revenues and government borrowing. As of January 2009 5,704 km of the intended 5,846 km has been 4 laned.

- Phase II: North-South and East-West corridors comprising national highways connecting four extreme points of the country. The North-South and East-West Corridor (NS-EW; 7,300 km) connecting Srinagar in the north to Kanyakumari in the south, including spur from Salem to Kochi (Via Coimbatore), and Silchar in the east to Porbandar in the west. Total length of the network is 7,300 km. As of January 2009, 42% of the project had been completed and 44% of the project work is currently at progress. It also includes Port connectivity and other projects — 1,157 km. The final completion date to February 28, 2009 at a cost of Rs350 billion (US$8 billion), with funding similar to Phase I.
- Phase III: The government recently approved NHDP-III to upgrade 12,109 km of national highways on a Build, Operate and Transfer (BOT) basis, which takes into account high-density traffic, connectivity of state capitals via NHDP Phase I and II, and connectivity to centres of economic importance. contracts have been awarded for a 2,075 km.
- Phase IV: The government is considering widening 20,000 km of highway that were not part of Phase I, II, or III. Phase IV will convert existing single lane highways into two lanes with paved shoulders. The plan will soon be presented to the government for approval.
- Phase V: As road traffic increases over time, a number of four lane highways will need to be upgraded/expanded to six lanes. The current plan calls for upgrade of about 5,000 km of four-lane roads, although the government has not yet identified the stretches.
- Phase VI: The government is working on constructing expressways that would connect major commercial and industrial townships. It has already identified 400 km of Vadodara (earlier Baroda)-Mumbai section that would connect to the existing Vadodara (earlier Baroda)-Ahmedabad section. The World Bank is studying this project. The project will be funded on BOT basis. The 334 km Expressway between Chennai—Bangalore (Now called Bengaluru) and 277 km Expressway between Kolkata—Dhanbad has been identified and feasibility study and DPR contract has been awarded by NHAI.

- Phase VII: This phase calls for improvements to city road networks by adding ring roads to enable easier connectivity with national highways to important cities. In addition, improvements will be made to stretches of national highways that require additional flyovers and bypasses given population and housing growth along the highways and increasing traffic. The government has not yet identified a firm investment plan for this phase.

Future Plans

The Indian Government has set ambitious plans for upgrading of the National Highways in a phased manner in the years to come. The details are as follows:

- 4-laning of 10,000 km (NHDP Phase-III) including 4,000 km that has been already approved. An accelerated road development programme for the North Eastern region.
- 2-laning with paved shoulders of 20,000 km of National Highways under NHDP Phase-IV.
- 6-laning of GQ and some other selected stretches covering 6,500 km under NHDP Phase-V.
- Development of 1,000 km of express ways under NHDP Phase-VI.
- Development of ring roads, bypasses, grade separators, service roads, etc. under NHDP Phase-VII.

Review of Urban Transportation in India

The establishment of State Transport Undertakings (STUs)1 in India in the 1960sand 1970s did an enormous service in linking towns and villages across the country, particularly in the western and southern parts. Even though the service mayleave much to be desired in terms of quality, the importance of STUs lies in the factthat, unlike in most other developing countries, one can connect to almost every village in India. Urban areas in India, which include a wide range of megacities, cities, and towns, are not all that fortunate in terms of intracity transportation. Transport in this context has been a victim of ignorance, neglect, and confusion. As far as the public transport system in Indian cities is concerned, dedicated citybus services are known to operate in 17 cities only and rail transit exists only in 4 out of 35 cities with population in excess of one million.

Transport demand in most Indian cities has increased substantially, due to increases in population as a result of both natural increase and migration from rural areas and smaller towns. Availability of motorized transport, increases in household income, and increases in commercial and industrial activities have further added to transport demand.

In many cases, demand has outstripped road capacity. Greater congestion and delays are widespread in Indian cities and indicate the seriousness of transport problems. A high level of pollution is another undesirable feature of overloaded streets. The transport crisis also takes a human toll. Statistics indicate that traffic accidents are a primary cause of accidental deaths in Indian cities. The main reasons for these problems are the prevailing imbalance in modal split, inadequate transport infrastructure, and its suboptimal use. Public transport systems have not been able to keep pace with the rapid and substantial increases in demand over the past few decades. Bus services in particular have deteriorated, and their relative output has been further reduced as passengers have turned topersonalized modes and intermediate public transport.

Individual cities cannot afford to cater only to private cars and two-wheelers. There must be a general recognition that without public transport cities would be even less viable. There is a need to encourage public transport instead of personal vehicles. This requires both an increase in quantity as well as quality of public transport and effective use of demand as well as supply-side management measures. People should also be encouraged to use nonmotorized transport and investments may be made to make it safer. Cities are the major contributors to economic growth, and movement in and between cities is crucial for improved quality of life.

Vehicular Growth and Modal Split

In 2002, 58.8 million vehicles were plying on Indian roads. According to statistics provided by the Ministry of Road Transport & Highways, Government of India, the annual rate of growth of motor vehicle population in India has been about 10 percent during the last decade. The basic problem is not the number of vehicles in the country but their concentration in a few selected cities, particularly in metropolitan cities (million plus). It is alarming to note that 32 percent of these vehicles are plying in metropolitan cities alone, which constitute about 11 percent of the total population. During the year 2000, more than 6.2 million vehicles were plying in megacities (Mumbai, Delhi, Kolkata, and Chennai) alone, which constitute more than 12.7 percent of all motor vehicles in the country.

Interestingly, Delhi, which contains 1.4 percent of the Indian population, accounts for nearly 7 percent of all motor vehicles in India.

Traffic composition in India is of a mixed nature. A wide variety of about a dozen types of both slow-and fast-moving vehicles exists. Two-wheelers and cars (including jeeps) account for more than 80 percent of the vehicle population inmost large cities. The share of buses is negligible inmost Indian cities as compared to personalized vehicles. For example, two-wheelers and cars together constitute more than 95 percent in Kanpur and 90 percent in both Hyderabad and Nagpur, whereas in these cities buses constitute 0.1, 0.3, and 0.8 percent, respectively.

Transport Infrastructure in Indian Cities

The area occupied by roads and streets in Class I cities (population more than 100,000) in India is only 16.1 percent of the total developed area, while the corresponding figure for the United States is 28.19 percent. Interestingly, even in Mumbai, the commercial capital of India, the percentage of space used for transportation is far less when viewed in comparison to its counterparts in the developed world. In general, the road space in Indian cities is grossly insufficient. To make the situation worse, most of the major roads and junctions in Indian cities are heavily encroached by parked vehicles, roadside hawkers, and pavement dwellers. As a consequence of these factors, the already deficient space for movement of vehicles is further reduced.

The present urban rail services in India are extremely limited. Only four cities (Mumbai, Delhi, Kolkata, and Chennai) are served by suburban rail systems. Rail services in these four main cities together carry more than 7 million trips per day. The Mumbai Suburban Rail System alone carries about 5.5 million trips per day. A few other cities also have limited suburban rail systems but they hardly meet the large transport demand existing in these cities.

A few metropolitan cities are served by well-organized bus services. Services are mostly run by publicly owned State Transport Undertakings (STUs). Private bus services operate mainly in Delhi and Kolkata. All passenger buses use the standard truck engine and chassis; hence, they are not economical for city use. There are virtually no buses in India specifically designed for urban conditions. Qualitatively, available urban mass transport services are overcrowded, unreliable, and involve long waiting periods. Overcrowding in the public transport system is more pronounced in large cities where buses, which are designed to carry 40 to 50 passengers generally, carry double the capacity during peak hours. As a result, there is a massive shift to personalized transport, especially two-wheelers, and proliferation of various types of intermediate public transport modes (three-wheeler auto-rickshaws and taxies).

Vehicular Emission, Congestion, and Road Safety Issues

The transport sector is the major contributor to air pollution in urban India. For example, 72 percent of air pollution in Delhi is caused by vehicular emission. According to studies by the Central Pollution Control Board (CPCB) of India, 76.2 percent of CO, 96.9 percent of hydrocarbons, and 48.6 percent of NOxare caused by emissions from the transport sector in Delhi. The ambient air pollution in terms of Suspended Particulate Matter (SPM) in all metropolitan cities in India exceeds the limit set by the World Health Organization (WHO) (Sharma and Mishra 1998). For example, in Kolkata, the average annual emission of SPM is 394 microgrammes per cubic meter, while the WHO standard is 75. With deteriorating levels of mass transport services and increasing use of personalized modes, vehicular emission has reached an alarming level in most Indian cities.

Indian cities also face severe traffic congestion. Growing traffic and limited road space have reduced peak-hour speeds to 5 to 10 kms per hour in the central areas of many major cities. This also leads to higher levels of vehicular emission. According to the Centre for Science and Environment (CSE), the quantity of all three major air pollutants (namely, CO, hydrocarbons, and nitrogen oxides) drastically increases with reduction in motor vehicle speeds. For example, at a speed of 75kmph, emission of CO is 6.4 gm/veh.-km, which increases by five times to 33.0 gm/veh.-km at a speed of 10 kmph. Similarly, emission of hydrocarbons, at the same speeds, increases by 4.8 times from 0.93 to 4.47 gm/veh.-km. Thus, prevalent traffic congestion in Indian cities, particularly during peak hours, not only increases the delay but also increases the pollution level.

India is also facing serious road accident problems. According to the Ministry of Road Transport & Highways, during 2001, nearly 80,000 people were killed in road accidents. In the last decade, road accidental deaths increased at a rate of 5 percent per year. Although annual rate of growth in road accidental deaths in Indian cities is a little less than 5 percent, these areas face serious road safety problems. For example, four Indian megacities constitute 5.4 percent of all road accident related fatalities, whereas only 4.4 percent of India's population lives in these areas. Analysis of data from a selected sample of cities shows that from 1990 to 1997, the number of fatalities is increasing at the rate of 4.1 percent per year—which is quite high by any standard. The accident severity index (number of fatalities per 100 accidents) was also found to be very high for all cities other than Ahmedabad, Bangalore, Kolkata, and Mumbai.

Policy Measures to Improve Urban Transportation in India

Focusing on Bus Transport

Passenger mobility in urban India relies heavily on its roads. Although rail-based transport services are available in a few megacities, they hardly play any role in meeting the transport demand in other million plus cities. Considering the financial health of various levels of governments (central, state, and local) and the investment required to improve the rail-based mass transport system, it is evident that bus transport will have to play a major role in providing passenger transport services in Indian cities in the future. It is amply clear that among the various modes of road based passenger transport, bus occupies less road space and causes less pollution per passenger-km than personalized modes. Therefore, urban transport plans should emphasize bus transport.

There is need for a great variety of bus transport services in Indian cities. Given the opportunity, people reveal widely divergent transport preferences, but in many places city authorities favour a basic standard of bus services. It is often thought to be in egalitarian to provide special services, such as guaranteed seats or express buses, in return for higher fares. In other words, variety is usually curbed. Government regulation and control have exacerbated the poor operational and financial performance of publicly owned urban transport undertakings, which are the main providers of bus transport services in Indian cities.

As cost of operation rises, transport systems come under financial pressure to raise fares, but politicians are under pressure to keep fares at existing levels. Unless the system is subsidized, it has to eliminate some of its less profitable or loss-making services. In a democracy, politicians are bound to yield to pressures from those whose services are threatened and to insist on maintaining money-losing operations. Due to this, transport undertakings find it difficult to raise their revenue sufficiently enough to meet the cost of operation. In addition, they have to provide concessional travel facilities to various groups, such as freedom fighters, journalists, students, besides paying a high level of different kinds of taxes. It is becoming increasingly difficult for loss-making urban transport undertakings to augment and manage their fleet, which in turn leads to poor operational performance and deterioration in quality of services.

With few exceptions, publicly owned urban transport undertakings in India operate at higher unit costs than comparable transport operations controlled by the private sector. Kolkata provides an opportunity to make a direct comparison between privately owned and publicly owned bus systems. Public buses are operated by the Calcutta State Transport Corporation (CSTC), with a fleet size of more than 1,250 buses and staffing ratio per operational bus of 11. CSTC has also been plagued by fare evasion estimated at more than 15 percent of revenue. As a result of low productivity and fare evasion, the system requires a huge subsidy since revenues cover less than half of the costs. On the other hand, there are 1,800 private buses in the city.

These buses are operated mainly by small companies or individual owners grouped into a number of route associations. Fares for private and public bus services are the same. Despite the similarity in fare rates, private operators have been able to survive financially without any subsidy. Their success is attributed to high levels of productivity, which are reflected in low staffing ratios and high fleet availability. Private bus operators in Kolkata, who hold almost two-thirds of the market, play a major role in meeting the demand and thus substantially reduce the financial burden on the state government. Furthermore, publicly owned urban transport undertakings often lack the flexibility of organization, the ability to hire and fire staff, or the financial discretion needed to adapt to changing conditions. In such circumstances, a policy that encourages private participation in the provision of bus transport services should be welcomed. There is an urgent need for restructuring of the public transport system in Indian cities to enhance both quantity as well as quality of services.

Enhancing Transport Coordination

There is an urgent need for a transportation system that is seamlessly integrated across all modes. The various modes of public transport, including intermediate public transport, have to work in tandem. They should complement rather than involve themselves in cutthroat competition. Presently, different agencies, independent of each other, are operating different services in Indian cities. For example, in Delhi, metro rail is operated by Delhi Metro Rail Corporation Ltd, suburban rail service by Northern Railway, bus transport service by Delhi Transport Corporation, and taxi and auto-rickshaw by private operators. There is a lack of coordination among these agencies. Since the ultimate objective is to provide an adequate and efficient transport system, there is a need to have a coordinating authority with the assigned role of coordinating the operations of various modes. This coordinating authority may be appointed by the central or state government and may have representatives from various stakeholders such as private taxi operators, bus operators, railways, and state government. The key objective should be to attain the integration of different modes of transport to improve the efficiency of service delivery and comfort for commuters. At the same time, a single-ticket system, where commuters can buy a transport ticket that is valid throughout the public transport network within the coordinating authority's jurisdiction, should also be developed and promoted.

Restraining the Use of Polluting Vehicles and Fuels

Most of the two-and three-wheelers in India operate with two-stroke engines, which emit a high volume of unburnt particles due to the incomplete combustion. Similarly, many new diesel cars have come up in the market, primarily because diesel is priced is far less than petrol in India. Government encourages this price differential mainly to help farmers and bus and truck operators. This price benefit is not meant to be available for personal cars. Although diesel cars emit less greenhouse gases, there are serious concerns about the public health effects of their particulate matter (PM) emissions in densely populated metropolitan cities. Government should use market-based instruments to promote cleaner technology and fuel. For example, a relatively high annual motor vehicle tax, which maybe increasing with the age of vehicle, can be imposed on two-stroke two-wheelers and all vehicles that are more than 10 years old. Similarly, cars that use diesel could be discouraged in million-plus cities by levying tax on diesel in those cities. Congestion pricing, parking fees, fuel taxes, and other measures could be used to restrain the use of all personalized modes. Emphasis should be on the use of market-based instruments as opposed to a command-and-control regime.

Demand-Side Management Measures

In general, Indian cities have not made much progress in implementing demand side management measures, such as congestion pricing and parking fees. Although policy measures that involve restraining the use of private cars and two-wheelers are likely to be unpopular, a gradualist approach of progressively introducing restraints on road use, while at the same time improving public transport, is more likely to lead to greater acceptance. Improved public transport and more efficient management of demand would help to combat the trend away from public transport vehicles and toward greater use of personalized modes.

Supply-Side Management Measures

Supply-side measures, such as one-way traffic, improvement of signals, traffic engineering improvements for road network and intersections, and bus priority lanes, should be introduced in all cities, especially in metropolitan cities, so that existing road capacity and road-user safety are increased. These may be considered short-term measures. Road infrastructure improvement measures, like new road alignments, hierarchy of roads, provision of service roads (e.g., bypasses, ring roads, bus bays, wide medians, intersection improvements, construction and repair of footpaths and roads, removal of encroachments, and good surface drainage) should also be introduced in million-plus cities. These can be considered medium term measures. Besides short-and medium-term measures, there is a need to have long-term measures as well, involving technology upgrades and the introduction of high-speed, high-capacity public transport systems particularly along high-density traffic corridors.

Transport systems are among the various factors affecting the quality of life and safety in a city. The urban transport situation in large cities in India is deteriorating. The deterioration is more prevalent in metropolitan cities where there is an excessive concentration of vehicles. Commuters in these cities are faced with acute road congestion, rising air pollution, and a high level of accident risk. These problems cannot be solved without a concise and cogent urban transport strategy. The main objective of such a strategy should be to provide and promote sustainable high-quality links for people by improving the efficiency and effectiveness of the city's transport systems. Policy should be designed in such a way as to reduce the need to travel by personalized modes and boost the public transport system. At the same time, demand-side as well as supply-side management measures should effectively be used. People should be encouraged to walk and cycle and government should support investments that make cycling and walking safer. Finally, there is a need to empower the Urban Local Bodies to raise finances and coordinate the activities of various agencies involved in the provision of transport infrastructure in urban areas.

9

International Transport

Air Traffic Control

Air traffic control (ATC) is a service provided by ground-based controllers who direct aircraft on the ground and in the air. The primary purpose of ATC systems worldwide is to *separate* aircraft to prevent collisions, to organize and expedite the flow of traffic, and to provide information and other support for pilots when able. In some countries, ATC may also play a security or defence role (as in the United States), or be run entirely by the military (as in Brazil).

Preventing collisions is referred to as separation, which is a term used to prevent aircraft from coming too close to each other by use of lateral, vertical and longitudinal separation minima; many aircraft now have collision avoidance systems installed to act as a backup to ATC observation and instructions. In addition to its primary function, the ATC can provide additional services such as providing information to pilots, weather and navigation information and NOTAMs (*NOtices To AirMen*).

In many countries, ATC services are provided throughout the majority of airspace, and its services are available to all users (private, military, and commercial). When controllers are responsible for separating some or all aircraft, such airspace is called "controlled airspace" in contrast to "uncontrolled airspace" where aircraft may fly without the use of the air traffic control system. Depending on the type of flight and the class of airspace, ATC may issue *instructions* that pilots are required to follow, or merely *flight information* (in some countries known as *advisories*) to assist pilots operating in the airspace. In all cases, however, the pilot in command has final responsibility for the safety of the flight, and may deviate from ATC instructions in an emergency.

In 1919, the International Commission for Air Navigation (ICAN) was created to develop General Rules for Air Traffic. Its rules and procedures were applied in most countries where aircraft operated. The United States did not sign the ICAN Convention, but later developed its own set of air traffic rules after passage of the Air Commerce Act of 1926. This legislation authorized the Department of Commerce to establish air traffic rules for the navigation, protection, and identification of aircraft, including rules as to safe altitudes of flight and rules for the prevention of collisions between vessels and aircraft. The first rules were brief and basic. For example, pilots were told not to begin their takeoff until there is no risk of collision with landing aircraft and until preceding aircraft are clear of the field. As traffic increased, some airport operators realized that such general rules were not enough to prevent collisions. They began to provide a form of air traffic control (ATC) based on visual signals. Early controllers, like Archie League (one of the first system's flagmen), stood on the field, waving flags to communicate with pilots.

As more aircraft were fitted for radio communication, radio-equipped airport traffic control towers began to replace the flagmen. In 1930, the first radio-equipped control tower in the United States began operating at the Cleveland Municipal Airport. By 1935, about 20 radio control towers were operating.

Increases in the number of flights created a need for ATC that was not just confined to airport areas but also extended out along the airways. In 1935, the principal airlines using the Chicago, Cleveland, and Newark airports agreed to coordinate the handling of airline traffic between those cities. In December, the first Airway Traffic Control Centre.opened at Newark, New Jersey. Additional centres at Chicago and Cleveland followed in 1936.

The early controllers tracked the position of planes using maps and blackboards and little boat-shaped weights that came to be called shrimp boats. They had no direct radio link with aircraft but used telephones to stay in touch with airline dispatchers, airway radio operators, and airport traffic controllers.

In July 1936, en route ATC became a federal responsibility and the first appropriation of $175,000 was made ($2,665,960 today). The Federal Government provided airway traffic control service, but local government authorities where the towers were located continued to operate those facilities.

In 1941, Congress appropriated funds for the Civil Aeronautics Administration (CAA) to construct and operate ATC towers, and soon the CAA began taking over operations at the first of these towers, with their number growing to 115 by 1944. In the postwar era, ATC at most airports was eventually to become a permanent federal responsibility. In response to wartime needs, the CAA also greatly expanded its en route air traffic control system.

The postwar years saw the beginning of a revolutionary development in ATC, the introduction of radar, a system that uses radio waves to detect distant objects. Originally developed by the British for military defence, this new technology allowed controllers to see the position of aircraft tracked on visual displays. In 1946, the CAA unveiled an experimental radar-equipped tower for control of civil flights. By 1952, the agency had begun its first routine use of radar for approach and departure control. Four years later, it placed a large order for long-range radars for use in en route ATC.

In 1960, the Federal Aviation Administration (FAA) began successful testing of a system under which flights in certain positive control areas were required to carry a radar beacon, called a transponder that identified the aircraft and helped to improve radar performance. Pilots in this airspace were also required to fly on instruments regardless of the weather and to remain in contact with controllers. Under these conditions, controllers were able to reduce the separation between aircraft by as much as half the standard distance.

For many years, pilots had negotiated a complicated maze of airways. In September 1964, the FAA instituted two layers of airways, one from 1,000 to 18,000 feet (305 to 5,486 meters) above ground level and the second from 18,000 to 45,000 feet (13,716 m) above mean sea level. It also standardized aircraft instrument settings and navigation checkpoints to reduce the controllers' workload. From 1965 to 1975, the FAA developed complex computer systems that would replace the plastic markers for tracking aircraft thereby modernizing the National Airspace System. Controllers could now view information sent by aircraft transponders to form alphanumeric symbols on a simulated three dimensional radar screen. The system allowed controllers to focus on providing separation by automating complex tasks.

The FAA established a Central Flow Control Facility in April 1970, to prevent clusters of congestion from disrupting the nationwide air traffic flow. This type of ATC became increasingly sophisticated and important, and in 1994, the FAA opened a new Air Traffic Control

System Command Centre with advanced equipment. In January 1982, the FAA unveiled the National Airspace System (NAS) Plan. The plan called for modernized flight service stations, more advanced systems for ATC, and improvements in ground-to-air surveillance and communication. Better computers and software were developed, air route traffic control centres were consolidated, and the number of flight service stations reduced. New Doppler Radars and better transponders complemented automatic, radio broadcasts of surface and flight conditions.

In July 1988, the FAA selected IBM to develop the new multi-billion-dollar Advanced Automation System (AAS) for the Nation's en route ATC centres. AAS would include controller workstations, called "sector suites," that would incorporate new display, communications and processing capabilities. The system had upgraded hardware enabling increased automation of complex tasks.

In December 1993, the FAA reviewed its order for the planned AAS. IBM was far behind schedule and had major cost overruns. In 1994 the FAA simplified its needs and picked new contractors. The revised modernization program continued under various project names. In 1999, controllers began their first use of an early version of the Standard Terminal Automation Replacement System, which included new displays and capabilities for approach control facilities. During the following year, FAA completed deployment of the Display System Replacement, providing more efficient workstations for en route controllers.

In 1994, the concept of Free Flight was introduced. It might eventually allow pilots to use on board instruments and electronics to maintain a safe distance between planes and to reduce their reliance on ground controllers. Full implementation of this concept would involve technology that made use of the Global Positioning System to help track the position of aircraft. In 1998, the FAA and industry began applying some of the early capabilities developed by the Free Flight program.

Current studies to upgrade ATC include the Communication, Navigation and Surveillance for Air Traffic Management System that relies on the most advanced aircraft transponder, a global navigation satellite system, and ultra-precise radar. Tests are underway to design new cockpit displays that will allow pilots to better control their aircraft by combining as many as 32 types of information about traffic, weather, and hazards.

Language

Pursuant to requirements of the International Civil Aviation Organization (ICAO), ATC operations are conducted either in the English language or the language used by the station on the ground. In practice, the native language for a region is normally used, however the English language must be used upon request.

Airport Control

The primary method of controlling the immediate airport environment is visual observation from the airport traffic control tower (ATCT). The ATCT is a tall, windowed structure located on the airport grounds. Aerodrome or Tower controllers are responsible for the separation and efficient movement of aircraft and vehicles operating on the taxiways and runways of the airport itself, and aircraft in the air near the airport, generally 2 to 5 nautical miles (3.7 to 9.2 km) depending on the airport procedures. Radar displays are also available to controllers at some airports. Controllers may use a radar system called Secondary Surveillance Radar for airborne traffic approaching and departing. These displays include a map of the area, the position of various aircraft, and data tags that include aircraft identification, speed, heading, and other information described in local procedures.

The areas of responsibility for ATCT controllers fall into three general operational disciplines; Local Control or Air Control, Ground Control, and Flight Data/Clearance Delivery—other categories, such as Apron Control or Ground Movement Planner, may exist at extremely busy airports. While each ATCT may have unique airport-specific procedures, such as multiple teams of controllers ('crews') at major or complex airports with multiple runways, the following provides a general concept of the delegation of responsibilities within the ATCT environment.

Ground Control

Ground Control (sometimes known as Ground Movement Control abbreviated to GMC or Surface Movement Control abbreviated to SMC) is responsible for the airport "movement" areas, as well as areas not released to the airlines or other users. This generally includes all taxiways, inactive runways, holding areas, and some transitional aprons or intersections where aircraft arrive, having vacated the runway or departure gate. Exact areas and control responsibilities are clearly defined in local documents and agreements at each airport. Any aircraft, vehicle, or person walking or working in these areas is required to have clearance from Ground Control. This is normally done via VHF/UHF radio, but there may be special cases where other processes are used.

Most aircraft and airside vehicles have radios. Aircraft or vehicles without radios must respond to ATC instructions via aviation light signals or else be led by vehicles with radios. People working on the airport surface normally have a communications link through which they can communicate with Ground Control, commonly either by handheld radio or even cell phone. Ground Control is vital to the smooth operation of the airport, because this position impacts the sequencing of departure aircraft, affecting the safety and efficiency of the airport's operation.

Some busier airports have Surface Movement Radar (SMR), such as, ASDE-3, AMASS or ASDE-X, designed to display aircraft and vehicles on the ground. These are used by Ground Control as an additional tool to control ground traffic, particularly at night or in poor visibility. There are a wide range of capabilities on these systems as they are being modernized. Older systems will display a map of the airport and the target. Newer systems include the capability to display higher quality mapping, radar target, data blocks, and safety alerts, and to interface with other systems such as digital flight strips.

Local Control or Air Control

Local Control (known to pilots as "Tower" or "Tower Control") is responsible for the active runway surfaces. Local Control clears aircraft for takeoff or landing, ensuring that prescribed runway separation will exist at all times. If Local Control detects any unsafe condition, a landing aircraft may be told to "go-around" and be re-sequenced into the landing pattern by the approach or terminal area controller.

Within the ATCT, a highly disciplined communications process between Local Control and Ground Control is an absolute necessity. Ground Control must request and gain approval from Local Control to cross any active runway with any aircraft or vehicle. Likewise, Local Control must ensure that Ground Control is aware of any operations that will impact the taxiways, and work with the approach radar controllers to create "holes" or "gaps" in the arrival traffic to allow taxiing traffic to cross runways and to allow departing aircraft to take off. Crew Resource Management (CRM) procedures are often used to ensure this communication process is efficient and clear, although this is not as prevalent as CRM for pilots.

Flight Data/Clearance Delivery

Clearance Delivery is the position that issues route clearances to aircraft, typically before they commence taxiing. These contain details of the route that the aircraft is expected to fly after departure. Clearance Delivery or, at busy airports, the Traffic Management Coordinator (TMC) will, if necessary, coordinate with the en route centre and national command centre or flow control to obtain releases for aircraft. Often, however, such releases are given automatically or are controlled by local agreements allowing "free-flow" departures. When weather or extremely high demand for a certain airport or airspace becomes a factor, there may be ground "stops" (or "slot delays") or re-routes may be necessary to ensure the system does not get overloaded.

The primary responsibility of Clearance Delivery is to ensure that the aircraft have the proper route and slot time. This information is also coordinated with the en route centre and Ground Control in order to ensure that the aircraft reaches the runway in time to meet the slot time provided by the command centre. At some airports, Clearance Delivery also plans aircraft pushbacks and engine starts, in which case it is known as the Ground Movement Planner (GMP): this position is particularly important at heavily congested airports to prevent taxiway and apron gridlock. Flight Data (which is routinely combined with Clearance Delivery) is the position that is responsible for ensuring that both controllers and pilots have the most current information: pertinent weather changes, outages, airport ground delays/ground stops, runway closures, etc. Flight Data may inform the pilots using a recorded continuous loop on a specific frequency known as the Automatic Terminal Information Service (ATIS).

Approach and Terminal Control

Many airports have a radar control facility that is associated with the airport. In most countries, this is referred to as *Terminal Control*; in the U.S., it is referred to as a TRACON (Terminal Radar Approach Control.) While every airport varies, terminal controllers usually handle traffic in a 30 to 50 nautical mile (56 to 93 km) radius from the airport. Where there are many busy airports close together, one consolidated TRACON may service all the airports. The airspace boundaries and altitudes assigned to a TRACON, which vary widely from airport to airport, are based on factors such as traffic flows, neighboring airports and terrain. A large and complex example is the London Terminal Control Centre which controls traffic for five main London airports up to 20,000 feet (6,100 m) and out to 100 nautical miles (190 km).

Terminal controllers are responsible for providing all ATC services within their airspace. Traffic flow is broadly divided into departures, arrivals, and overflights. As aircraft move in and out of the terminal airspace, they are handed off to the next appropriate control facility (a control tower, an en-route control facility, or a bordering terminal or approach control). Terminal control is responsible for ensuring that aircraft are at an appropriate altitude when they are handed off, and that aircraft arrive at a suitable rate for landing.

Not all airports have a radar approach or terminal control available. In this case, the en-route centre or a neighboring terminal or approach control may co-ordinate directly with the tower on the airport and vector inbound aircraft to a position from where they can land visually. At some of these airports, the tower may provide a non-radar procedural approach service to arriving aircraft handed over from a radar unit before they are visual to land. Some units also have a dedicated approach unit which can provide the procedural approach service either all the time or for any periods of radar outage for any reason.

En-route, Centre, or Area Control

ATC provides services to aircraft in flight between airports as well. Pilots fly under one of two sets of rules for separation: Visual Flight Rules (VFR) or Instrument Flight Rules (IFR). Air traffic controllers have different responsibilities to aircraft operating under the different sets of rules. While IFR flights are under positive control, in the US VFR pilots can request flight following, which provides traffic advisory services on a time permitting basis and may also provide assistance in avoiding areas of weather and flight restrictions. In the UK, a pilot can request for "Deconfliction Service", which is similar to flight following. En-route air traffic controllers issue clearances and instructions for airborne aircraft, and pilots are required to comply with these instructions. En-route controllers also provide air traffic control services to many smaller airports around the country, including clearance off of the ground and clearance for approach to an airport. Controllers adhere to a set of separation standards that define the minimum distance allowed between aircraft. These distances vary depending on the equipment and procedures used in providing ATC services.

General Characteristics

En-route air traffic controllers work in facilities called Area Control Centres, each of which is commonly referred to as a "Centre". The

United States uses the equivalent term Air Route Traffic Control Centre (ARTCC). Each centre is responsible for many thousands of square miles of airspace (known as a Flight Information Region) and for the airports within that airspace. Centres control IFR aircraft from the time they depart from an airport or terminal area's airspace to the time they arrive at another airport or terminal area's airspace. Centres may also "pick up" VFR aircraft that are already airborne and integrate them into the IFR system. These aircraft must, however, remain VFR until the Centre provides a clearance.

Centre controllers are responsible for climbing the aircraft to their requested altitude while, at the same time, ensuring that the aircraft is properly separated from all other aircraft in the immediate area. Additionally, the aircraft must be placed in a flow consistent with the aircraft's route of flight. This effort is complicated by crossing traffic, severe weather, special missions that require large airspace allocations, and traffic density. When the aircraft approaches its destination, the centre is responsible for meeting altitude restrictions by specific points, as well as providing many destination airports with a traffic flow, which prohibits all of the arrivals being "bunched together". These "flow restrictions" often begin in the middle of the route, as controllers will position aircraft landing in the same destination so that when the aircraft are close to their destination they are sequenced.

As an aircraft reaches the boundary of a Centre's control area it is "handed off" or "handed over" to the next Area Control Centre. In some cases this "hand-off" process involves a transfer of identification and details between controllers so that air traffic control services can be provided in a seamless manner; in other cases local agreements may allow "silent handovers" such that the receiving centre does not require any co-ordination if traffic is presented in an agreed manner. After the hand-off, the aircraft is given a frequency change and begins talking to the next controller. This process continues until the aircraft is handed off to a terminal controller ("approach").

Radar Coverage

Since centres control a large airspace area, they will typically use long range radar that has the capability, at higher altitudes, to see aircraft within 200 nautical miles (370 km) of the radar antenna. They may also use TRACON radar data to control when it provides a better "picture" of the traffic or when it can fill in a portion of the area not covered by the long range radar.

In the U.S. system, at higher altitudes, over 90% of the U.S. airspace is covered by radar and often by multiple radar systems; however, coverage may be inconsistent at lower altitudes used by unpressurized aircraft due to high terrain or distance from radar facilities. A centre may require numerous radar systems to cover the airspace assigned to them, and may also rely on pilot position reports from aircraft flying below the floor of radar coverage. This results in a large amount of data being available to the controller. To address this, automation systems have been designed that consolidate the radar data for the controller. This consolidation includes eliminating duplicate radar returns, ensuring the best radar for each geographical area is providing the data, and displaying the data in an effective format.

Centres also exercise control over traffic travelling over the world's ocean areas. These areas are also FIRs. Because there are no radar systems available for oceanic control, oceanic controllers provide ATC services using procedural control. These procedures use aircraft position reports, time, altitude, distance, and speed to ensure separation. Controllers record information on flight progress strips and in specially developed oceanic computer systems as aircraft report positions. This process requires that aircraft be separated by greater distances, which reduces the overall capacity for any given route.

Some Air Navigation Service Providers (e.g. Airservices Australia, The Federal Aviation Administration, NAVCANADA, etc.) have implemented Automatic Dependent Surveillance-Broadcast (ADS-B) as part of their surveillance capability. This new technology reverses the radar concept. Instead of radar "finding" a target by interrogating the transponder, the ADS-equipped aircraft sends a position report as determined by the navigation equipment on board the aircraft. Normally, ADS operates in the "contract" mode where the aircraft reports a position, automatically or initiated by the pilot, based on a predetermined time interval. It is also possible for controllers to request more frequent reports to more quickly establish aircraft position for specific reasons. However, since the cost for each report is charged by the ADS service providers to the company operating the aircraft, more frequent reports are not commonly requested except in emergency situations. ADS is significant because it can be used where it is not possible to locate the infrastructure for a radar system (e.g. over water). Computerized radar displays are now being designed to accept ADS inputs as part of the display. This technology is currently used in portions of the North Atlantic and the Pacific by a variety of states who share responsibility for the control of this airspace.

Precision approach radars are commonly used by military controllers of airforces of several countries, to assist the Pilot in final phases of landing in places where Instrument Landing System and other sophisticated air borne equipements are unavailable to assist the pilots in marginal or *near zero visibility* conditions. This procedure is also called Talkdowns.

Flight Traffic Mapping

The mapping of flights in real-time is based on the air traffic control system. In 1991, data on the location of aircraft was made available by the Federal Aviation Administration to the airline industry. The National Business Aviation Association (NBAA), the General Aviation Manufacturers Association, the Aircraft Owners & Pilots Association, the Helicopter Association International, and the National Air Transportation Association petitioned the FAA to make ASDI information available on a "need-to-know" basis. Subsequently, NBAA advocated the broad-scale dissemination of air traffic data. The Aircraft Situational Display to Industry (ASDI) system now conveys up-to-date flight information to the airline industry and the public. Some companies that distribute ASDI information are FlightExplorer, FlightView, and FlyteComm. Each company maintains a website that provides free updated information to the public on flight status. Stand-alone programs are also available for displaying the geographic location of airborne IFR (Instrument Flight Rules) air traffic anywhere in the FAA air traffic system. Positions are reported for both commercial and general aviation traffic. The programs can overlay air traffic with a wide selection of maps such as, geo-political boundaries, air traffic control centre boundaries, high altitude jet routes, satellite cloud and radar imagery.

Problems

Traffic

The day-to-day problems faced by the air traffic control system are primarily related to the volume of air traffic demand placed on the system and weather. Several factors dictate the amount of traffic that can land at an airport in a given amount of time. Each landing aircraft must touch down, slow, and exit the runway before the next crosses the beginning of the runway. This process requires at least one and up to four minutes for each aircraft. Allowing for departures between arrivals, each runway can thus handle about 30 arrivals per hour. A large airport with two arrival runways can handle about 60 arrivals per hour in good weather. Problems begin when airlines schedule more arrivals into an airport than can be physically handled, or when delays elsewhere cause groups of aircraft that would otherwise be separated in time to arrive simultaneously. Aircraft must then be delayed in the air by holding over specified locations until they may be safely sequenced to the runway. Up until the 1990s, holding, which has significant environmental and cost implications, was a routine occurrence at many airports. Advances in computers now allow the sequencing of planes hours in advance. Thus, planes may be delayed before they even take off (by being given a "slot"), or may reduce speed in flight and proceed more slowly thus significantly reducing the amount of holding.

Weather

Beyond runway capacity issues, weather is a major factor in traffic capacity. Rain, ice or snow on the runway cause landing aircraft to take longer to slow and exit, thus reducing the safe arrival rate and requiring more space between landing aircraft. Fog also requires a decrease in the landing rate. These, in turn, increase airborne delay for holding aircraft. If more aircraft are scheduled than can be safely and efficiently held in the air, a ground delay program may be established, delaying aircraft on the ground before departure due to conditions at the arrival airport.

In Area Control Centres, a major weather problem is thunderstorms, which present a variety of hazards to aircraft. Aircraft will deviate around storms, reducing the capacity of the en-route system by requiring more space per aircraft, or causing congestion as many aircraft try to move through a single hole in a line of thunderstorms. Occasionally weather considerations cause delays to aircraft prior to their departure as routes are closed by thunderstorms.

Much money has been spent on creating software to streamline this process. However, at some ACCs, air traffic controllers still record data for each flight on strips of paper and personally coordinate their paths. In newer sites, these flight progress strips have been replaced by electronic data presented on computer screens. As new equipment is brought in, more and more sites are upgrading away from paper flight strips.

Call Signs

A prerequisite to safe air traffic separation is the assignment and use of distinctive call signs. These are permanently allocated by ICAO (pronounced “ai-kay-oh”) on request usually to scheduled flights and some air forces for military flights. They are written callsigns with 3-letter combination like KLM, AAL, SWA, BAW, DLH followed by the flight number, like AAL872, BAW018. As such they appear on flight plans and ATC radar labels. There are also the audio or Radio-telephony callsigns used on the radio contact between pilots and Air Traffic Control not always identical with the written ones. For example BAW stands for British Airways but on the radio you will only hear the word Speedbird instead. By default, the callsign for any other flight is the registration number (tail number) of the aircraft, such as “N12345” or “C-GABC”. The term tail number is because a registration number is usually painted somewhere on the tail of a plane, yet this is not a rule. Registration numbers may appear on the engines, anywhere on the fuselage, and often on the wings. The short Radio-telephony callsigns for these tail numbers is the last 3 letters only like ABC spoken Alpha-Bravo-Charlie for C-GABC or the last 3 numbers like 345 spoken as three-four-five for N12345. In the United States the abbreviation of callsigns is required to be a prefix (such as aircraft type, aircraft manufacturer, or first letter of registration) followed by the last three characters of the callsign. This abbreviation is only allowed after communications has been established in each sector.

The flight number part is decided by the aircraft operator. In this arrangement, an identical call sign might well be used for the same scheduled journey each day it is operated, even if the departure time varies a little across different days of the week. The call sign of the return flight often differs only by the final digit from the outbound flight. Generally, airline flight numbers are even if eastbound, and odd if westbound. In order to reduce the possibility of two callsigns on one frequency at any time sounding too similar, a number of airlines, particularly in Europe, have started using alphanumeric callsigns that are not based on flight numbers. For example DLH23LG, spoken as lufthansa-two-tree-lima-golf. Additionally it is the right of the air traffic controller to change the ‘audio’ callsign for the period the flight is in his sector if there is a risk of confusion, usually choosing the tail number instead.

Before around 1980 International Air Transport Association (IATA) and ICAO were using the same 2-letter callsigns. Due to the larger number of new airlines after deregulation ICAO established the 3-letter callsigns as mentioned above. The IATA callsigns are currently used in aerodromes on the announcement tables but never used any longer in Air Traffic Control. For example, AA is the IATA callsign for American Airlines — ATC equivalent AAL. Other examples include LY/ELY for El Al, DL/DAL for Delta Air Lines, LH/DLH for Lufthansa etc.

Technology

Many technologies are used in air traffic control systems. Primary and secondary radar are used to enhance a controller's situation awareness within his assigned airspace — all types of aircraft send back primary echoes of varying sizes to controllers' screens as radar energy is bounced off their skins, and transponder-equipped aircraft reply to secondary radar interrogations by giving an ID (Mode A), an altitude (Mode C) and/or a unique callsign (Mode S). Certain types of weather may also register on the radar screen.

These inputs, added to data from other radars, are correlated to build the air situation. Some basic processing occurs on the radar tracks, such as calculating ground speed and magnetic headings.

Usually, a Flight Data Processing System manages all the flight plan related data, incorporating-in a low or high degree-the information of the track once the correlation between them (flight plan and track) is established. All this information is distributed to modern operational display systems, making it available to controllers.

The FAA has spent over USD$3 billion on software, but a fully-automated system is still over the horizon. In 2002 the UK brought a new area control centre into service at Swanwick, in Hampshire, relieving a busy suburban centre at West Drayton in Middlesex, north of London Heathrow Airport. Software from Lockheed-Martin predominates at Swanwick. However, Swanwick was initially troubled by software and communications problems causing delays and occasional shutdowns. Some tools are available in different domains to help the controller further:

- Flight Data Processing Systems: this is the system (usually one per Centre) that processes all the information related to the Flight (the Flight Plan), typically in the time horizon from Gate to gate (airport departure/arrival gates). It uses such processed information to invoke other Flight Plan related tools (such as e.g. MTCD), and distributes such processed information to all the stakeholders (Air Traffic Controllers, collateral Centres, Airports, etc.).
- Short Term Conflict Alert (STCA) that checks possible conflicting trajectories in a time horizon of about 2 or 3 minutes

(or even less in approach context-35 seconds in the French Roissy & Orly approach centres) and alerts the controller prior to the loss of separation. The algorithms used may also provide in some systems a possible vectoring solution, that is, the manner in which to turn, descend, or climb the aircraft in order to avoid infringing the minimum safety distance or altitude clearance.

- Minimum Safe Altitude Warning (MSAW): a tool that alerts the controller if an aircraft appears to be flying too low to the ground or will impact terrain based on its current altitude and heading.
- System Coordination (SYSCO) to enable controller to negotiate the release of flights from one sector to another.
- Area Penetration Warning (APW) to inform a controller that a flight will penetrate a restricted area.
- Arrival and Departure Manager to help sequence the takeoff and landing of aircraft.
    - The Departure Manager (DMAN): A system aid for the ATC at airports, that calculates a planned departure flow with the goal to maintain an optimal throughput at the runway, reduce queuing at holding point and distribute the information to various stakeholders at the airport (i.e. the airline, ground handling and Air Traffic Control (ATC)).
    - The Arrival Manager (AMAN): A system aid for the ATC at airports, that calculates a planned Arrival flow with the goal to maintain an optimal throughput at the runway, reduce arrival queuing and distribute the information to various stakeholders.
    - passive Final Approach Spacing Tool (pFAST), a CTAS tool, provides runway assignment and sequence number advisories to terminal controllers to improve the arrival rate at congested airports. pFAST was deployed and operational at five US TRACONs before being cancelled. NASA research included an Active FAST capability that also provided vector and speed advisories to implement the runway and sequence advisories.
- Converging Runway Display Aid (CRDA) enables Approach controllers to run two final approaches that intersect and make sure that go arounds are minimized
- Centre TRACON Automation System (CTAS) is a suite of human centered decision support tools developed by NASA

Ames Research Centre. Several of the CTAS tools have been field tested and transitioned to the FAA for operational evaluation and use. Some of the CTAS tools are: Traffic Management Advisor (TMA), passive Final Approach Spacing Tool (pFAST), Collaborative Arrival Planning (CAP), Direct-To (D2), En Route Descent Advisor (EDA) and Multi Centre TMA.

- Traffic Management Advisor (TMA), a CTAS tool, is an en route decision support tool that automates time based metering solutions to provide an upper limit of aircraft to a TRACON from the Centre over a set period of time. Schedules are determined that will not exceed the specified arrival rate and controllers use the scheduled times to provide the appropriate delay to arrivals while in the en route domain. This results in an overall reduction in en route delays and also moves the delays to more efficient airspace (higher altitudes) than occur if holding near the TRACON boundary is required to not overload the TRACON controllers. TMA is operational at most en route air route traffic control centres (ARTCCs) and continues to be enhanced to address more complex traffic situations (e.g. Adjacent Centre Metering (ACM) and En Route Departure Capability (EDC))
- MTCD & URET
  - o In the US, User Request Evaluation Tool (URET) takes paper strips out of the equation for En Route controllers at ARTCCs by providing a display that shows all aircraft that are either in or currently routed into the sector.
  - o In Europe, several MTCD tools are available: iFACTS (NATS), ERATO (DSNA fr:DSNA), VAFORIT (DFS), New FDPS (MASUAC). The SESAR Programme should soon launch new MTCD concepts.

URET and MTCD provide conflict advisories up to 30 minutes in advance and have a suite of assistance tools that assist in evaluating resolution options and pilot requests.

- Mode S: provides a data downlink of flight parameters via Secondary Surveillance Radars allowing radar processing systems and therefore controllers to see various data on a flight, including airframe unique id (24-bits encoded), indicated airspeed and flight director selected level, amongst others.
- CPDLC: Controller Pilot Data Link Communications — allows digital messages to be sent between controllers and pilots, avoiding the need to use radiotelephony. It is especially useful

in areas where difficult-to-use HF radiotelephony was previously used for communication with aircraft, e.g. oceans. This is currently in use in various parts of the world including the Atlantic and Pacific oceans.

- ADS-B: Automatic Dependent Surveillance Broadcast—provides a data downlink of various flight parameters to air traffic control systems via the Transponder (1090 MHz) and reception of those data by other aircraft in the vicinity. The most important is the aircraft's latitude, longitude and level: such data can be utilized to create a radar-like display of aircraft for controllers and thus allows a form of pseudo-radar control to be done in areas where the installation of radar is either prohibitive on the grounds of low traffic levels, or technically not feasible (e.g. oceans). This is currently in use in Australia, Canada and parts of the Pacific Ocean and Alaska.
- Screen Content Recording: Hardware or software based recording function which is part of most modern Automation System and that captures the screen content shown to of the ATCO. Such recordings are used for a later replay together with audio recording for investigations and post event analysis.

Transportation in Developing Countries

In rural areas and small cities of China and India, millions of small locally made three-and four-wheel "rural vehicles" are proliferating. In China, the vehicles are banned in large cities because of their slow speed and high emissions, but even so rural vehicle sales in China outnumber those of conventional cars and trucks. These vehicles, which cost anywhere from $400 to $4,000 each, are the heart of millions of small businesses, transporting farm products, construction materials, and locally manufactured products. They also serve as the principal mode of motorized travel in rural areas.

Motorization is accelerating even more rapidly in cities. Personal vehicles, from scooters to large company cars, are improving access to goods, services, and activities, including an expanded array of job and educational opportunities. They provide unmatched flexibility, convenience, and freedom. For many individuals, vehicles are desirable as a secure and private means of travel, and as status symbols. For businesses, they are a means of increasing productivity.

But personal motorization also imposes enormous costs, especially in cities. The well-known litany includes air and noise pollution, neighborhood fragmentation, and high energy use. Motorized transport is the largest consumer of the world's petroleum, making it central to international concerns over energy security and political stability in volatile regions. China is now the second largest importer of oil in the world, although its vehicle ownership rates are but one-fiftieth of the US's. The developing world is an increasing source of greenhouse gas (GHG) emissions, which are rising faster in transportation than in any other sector. Developing cities and countries are in a quandary. How can they accommodate the intense desire for personal mobility while mitigating the heavy economic, environmental, and social costs of motorization?

For countries such as India and China, which look to automotive manufacturing as a pillar of economic development, the dilemma is extreme.

Transportation Planning Problems in Developing Countries

The importance of planning in transportation is obvious. As residents of Istanbul and Turkey, we can easily observe the negative effects of unplanned transportation, and see that ineffective planning, with regard to its concepts and instruments does not reduce the problems at all.

The most striking fact of the last fifteen years of the transportation field in Turkey is that 1700 kilometres of motorways were built by spending 16 billion US dollars. On the other hand, at the stage where decisions were being made for these motorways, the "1983-1993 National Transportation Plan" was in operation and revisions were being conducted after its first three years. Decisions of building 1200 kilometres of motorways were made in spite of the fact that no new motorways were suggested in this national plan. On the contrary, a high-speed railway was suggested, and a quarter of the construction was complete when it was abandoned in favour of the new motorways. The National Plan was abandoned in this manner and without any reason and without any explanation at all. Today, none of these motorways use more than 10 % of their capacities. Another issue simultaneous with the motorway decision was the construction of new airports using the slogan "a new airport to each province". These airports were recently closed after about ten years of operation during which only a few planes used them weekly.

Another fact is the planning cycle formed by continuously repeating plan and project making. No feasibility studies were made for the motorways mentioned above, and even some had their projects made by the contractor firms after the beginning of the construction.

Studies and projects continued to be made many times for the abandoned high-speed railway. Another exaggerated example is the city of Bursa. It had its transportation master plan made 3 times in 10 years. Some of the examples are unplanned developments resulting from the lack of or a misunderstanding of the planning concept. The others are developments against plans under operation. The lack of a planning tradition causes diffidence against the planning concept and reduces the effects of the efforts made in favour of planned development. Planning approaches are criticised also in developed countries, where there is a planning tradition and where planned development is under operation. Some of the criticism centreson methodology and instruments (Genton, 1971; Wachs, 1985). Some say that instrumental approach should be replaced by the communicative approach (Willson, 2001). Most of the criticism focuses on demand predictions, which lie in the centre of the classical planning processes.

Talvitie (1997), who approaches the subject from a philosophical direction and bases his ideas on economical and psychoanalytic theories, believes that transportation and societal planning is extensive far beyond the individuals' economical behaviour. He proposes that the utility function should be should be expanded beyond the limits of economical behaviour. He stresses that the following three questions should be positively answered because of the important role of demand prediction in planning:

1. Can socio-demography, land use and travel demands be forecast as a function of observable variables?
2. Can stable goals and plans be formulated, satisfying both the goals and predictions?
3. Has a tractable process been devised for implementing the plan?

The only possible answer for all of these questions is stated to be "no". Also added is the fact that the effect of survey errors and errors resulting from unknown and unobservable variables is high regarding the results. Thus, these subjects should be...Planning problems are deeper in the developing countries. Without the necessary tradition and the past experiences, the planning efforts cannot yield the best results in a short period of time. First of all, the planning concept should be adopted and assimilated. The question is not "why planning?" This stage is over. The questions of "what kind of plan" and what kind of application" are waiting to be answered. This paper consists of some ideas on the answers to these questions, with regard to the developing countries.

Level of Development and Transportation

The relationship between the level of development and transportation can be examined from different points of view. Certainly the most dominant indicator, among others, is the National income. Low national income results in poor transportation infrastructure. Therefore, big investments continue to be necessary in the developing countries. The flexibility of transport demand against the national income is higher than 1. For example, as the increase in national income in Turkey in the 25-year period between the years 1970 and 1995, had an annual average of 4.31%, there were increases of 5.28% in passenger-km and 5.95% ton-km values. In this case, a 1% increase in national income ignites an increase of 1.23% in passenger demand and 1.38% in freight demand. These figures are 2.34% for passengers and 0.90% for freight respectively in the 15 EU countries (European Commision, 1997). The increasing trend in the passenger transport is an expected development.

The relationship between the level of development and the level of car ownership is also obvious. The rapid increase of car ownership in the developed countries, and the fact that today it corresponds nearly to one car per two persons, has brought forward the problem of traffic congestion. On the other hand, it has also been understood through the years that the whole problem cannot be solved by just solving the congestion problem. Although the level of car ownership is relatively lower in the developing countries, over the years traffic congestion has been accepted as the primary transportation problem indicator. Thus, the heart of the matter is missed and only the symptom is taken into consideration instead of the real problem. This traffic congestion does not exist only in the rush hours, but it continues throughout the day in cities like Istanbul. Actually, this problem has become a common syndrome for most of the big cities in developing countries (Bovy, 1976).

The specific reasons of traffic congestion in Istanbul and other cities in developing countries can be summarised as follows:

- The urban texture is not appropriate for cars.
- Car owners spare a relatively big amount of time budget for transportation; ie they tend to use cars many times and for longer periods. Thus, the rate of car usage is high.
- Traffic management is inadequate.
- Disobedience to the traffic rules is widespread.
- Public transport is inadequate.

Actually, most of the measures taken against congestion are meaningless because they lack permanency and do not solve the non-car-owning majority's problems. Thus, it can easily be said that the real problem waiting to be solved in the developing countries, is public transportation, which the vast majority depends on. The average age of the population and the present rate of mobility are low (eg, daily mobility in Istanbul is 1.0, whereas it is between 2.0 and 3.5 in the developed countries). This fact means that in the future, the rate of the population depending on public transport will increase.

Planning Based on Development Conditions

Transport plan can be expressed as a process of activities formed up for the purpose of meeting the cost of transport needs arising from the socio-economic activities in a specific region at an acceptable level of service and along with its external costs. Making correct plans for the desired development is a subject of long-term studies. The developments taken place in the last 50 years are mostly structured around the developed economical and social systems of the USA and European countries. As far as the expressed structural features and facilities are concerned, it becomes even more essential to investigate this subject deeply for developing countries.

In the countries, where the current planning methodologies have been developed, there is astable balance in terms of economic and social aspects. The developments are, to a great extent, take place in the desired way. For this reason, efforts for creating the supply to meet the increasing transport demand with respect to increasing population, growing economy and search for quality have gained importance. This situation results in a requirement for the prediction of the demand, creation of different alternatives for supply and selecting the optimum. In the prediction of the demand that lies in the centre of the planning process, the assumption that the former trends in the selection of transport modes would be the same in the future, might be valid in the steady state structures of these countries. Furthermore, the data required for forecasting and assessments are usually reliable and adequate. Since the transportation infrastructure in these countries have almost been completed, investments of great importance may not be necessary.

However, in the developing countries, in terms of economy and the geographical distribution of socio-economic activities, definition of predictable trends extending from the past to the future is mostly impossible. The developments might take place in the forms of leaps rather than linear improvements. In this case, investments may be much more expensive depending on demand predictions based on unrealistic assumptions.

In some big cities, as a result of inconsistencies in land use and rapid increase of the population, possibility of creating supply capacity may disappear in practice. In this case, validity of common planning methodology might become controversial and land use planning may become primary.

Planning in Developing Countries

With respect to the explanations given above, might the planning approach and its methodology, widely applied in the USA and European countries be valid outside these countries, eg Turkey? The answer is unlikely to be positive. Because the core of this approach mentioned above, a different formulation should be assessed for the developing countries. For example, the development path in Turkey first needs to be stopped and then altered; because the highways, which carry 95% of the passengers and 93% of the freight, have almost taken the whole burden of transportation.

The railways seem to be approaching an end. Despite being inside a lively sector, the airways are a transport mode that has limited opportunities for growing. Although every geographical opportunity exists for it, marine transportation cannot be developed. And of course, there is the fact that around 7000-10000 people die in traffic accidents each year.

The share of highways in the city of Istanbul is around 92% despite the effort put in developing the rail systems. The time lost in urban transportation and its stress causes production losses and bring along negative effects on peoples' emotional state. Nationwide transportation in Turkey and urban transportation in Istanbul is in such a state that because of this imbalanced development, it can no longer reach a healthy structure. So, instead of supporting the present trend, it should be halted and redirected in order to solve the problems. In short, it can be said that a "passive" approach, which predicts the future through present trends and developments instead of directing it, can be valid for the developed countries. "Active" approaches that are directive in this sense are necessary for the developing countries. Accordingly, the planning process should have an accomplished structure based on finding the most consistent and valid solution on economical and social development as well as land use, instead of satisfying the demand for the foreseeable future.

The Decisive Importance of Public Transport

Public transport forms the skeleton of the transportation systems in the big cities. As stated before, especially in the developing countries the rate of the population depending on public transport is quite high. The state of public transport in the cities of the developing countries differs widely from those of the developed countries. This difference brings different conflicts to the surface. Public transport in the cities of the developed countries can be considered adequate, at least when compared with the developing countries. However, this adequacy is not sufficient enough to attract the car-users to public transport.

In the developing countries, urban demand on public transport is higher. Yet, there is no existent public transport capacity that will attract the car-users. In this situation, congestion created by the relatively low car ownership comes to forefront and the unfair and illogical sharing of the transport facilities establish the main conflict. For example, the cars that use 70% of the roads carry only 20% of the travels, while the buses, which use 4% of the roads, serve 35%. The planning approach foresees the system development according to the past trends in private car demand. Since public transport is weak, this conflict is abided to. Bus ways become necessary on certain axes and at certain stages of the demand. However, the priority is given to the cars and the share of the road that are necessary for them are determined first; the remaining portion is then allocated to buses. With this approach, the problem cannot be solved at all. On the other hand, at demand values, where rail systems become obligatory, feasibility studies mostly turn out to give negative results, not only financially but also economically.

The reason for this is that the value of time, which has wights of up to 80% among the utilities, stay at low levels as a function of national income. Economical values of fatalities and injuries in traffic accidents are disturbing even ethically and the decrease in fatalities and injuries do not make a positive contribution to the feasibility results. The main conflict in the developing countries and their cities is that the capacities (bus ways or especially rail systems) which are necessitated by the demand cannot be created or there is an important delay in this process. Two natural results of this are:

- Required transportation support is not given for a healthy urban development; furthermore, hope for a healthy city disappears.
- The delay compensation resulting from the delay in constructing the high-capacity transportation system rises rapidly and surpasses the investment cost.

For example, this situation is present in Istanbul. Costs of investments such as the subway are increasing rapidly while alternative costs as operating costs reach quite high values. The results of an old study on the public transport systems, and especially rail systems in the major cities of the developing countries are quite interesting. In this study financed by the OECD, survey results from 21 different cities including Istanbul, which have rail systems or have rail system projects under operation, are evaluated. The total income curves (ie. population x income per capita) in the graph indicate that all of the cities having an income of over 15 million USD have subway systems; whereas, most cities having an income of lower than 5 million USD do not have them. This means that even cities with populations over 10 million do not possess rail systems.

The most important reason stated for building a subway system is the improvement of public transportation. This is followed by the requirement for solving or reducing traffic congestion, which should be considered as a false concept, as mentioned above. Another aim is to support land use planning policies. The main reason stated for Istanbul is this latter one. The reasons of the problems in rail systems are also examined in this study. The main reason is found to be the mistakes made in the planning stage. According to this, no feasibility studies were conducted for some of the subway systems.

Those that were conducted are far from being adequate in both quality and scope. Significant deviations are observed in demand and cost predictions compared to reality. Timetable and price matching with other systems is usually neglected. Errors can be made in route selection. Some of these errors can be of great extent and unrecoverable.

Planning Methodology

For the developing countries, transportation planning has become a methodological problem, or even further, a modelling problem; and has become equivalent to the modelit self. Reflection of the approach's passive behaviour on the methodology is a natural case. Thus, it may not be very useful to go into the details of the methodology. Meaningful results can be obtained though, by mentioning certain points.

- The assumption that relationships based on measurements made in a time interval will not change in the prediction period will bring the model into a static state.
- There are many hidden and arbitrary assumptions in its structure.
- With demand production, travel decision is made before the idea of where, how and why to go. This means that it is not affected by the transportation system.

- It is assumed that the mobility of car owners will continue. No justice or fairness is considered in mobility between different social groups.
- There is the possibility of a well-calibrated model giving wrong results for movements between certain zones.
- There is no possibility for considering preferences arising from the inadequacy of the public transport system.
- Effects in the distant future are underrated by the actualisation technique.

For example, the reductions in traffic accidents cannot affect the result because of high actualisation.

Transportation Management and Financing

Since transportation management has many authorities in the developing countries, its ability of coordination is either quite low or there is no such ability at all. Thus:

- Even though plans are made, they cannot be applied with stability and without compensation. The planners try to operate the model without taking this structure into consideration.
- The investments do not always satisfy the primary needs. They sometimes cause big amounts of money to be wasted.
- The present potential cannot be utilised effectively.
- The issue of financing is neglected in the planning stage because management is chaotic from planning to investment, operation and supervising. The planned investments lose the chance of being realised. The plan cannot become an important document from the beginning. Actually financing is primary issue for the developing countries. Isolating it from the management's effectiveness cannot solve this issue.

Bibliography

Allen, Kathleen M. S., Stanton W. Green: *Interpreting Space: GIS and Archaeology*, London and New York: Taylor & Francis, 1990.

Balshaw, Maria, and Liam Kennedy: *Urban Space and Representation.* Sterling, VA: Pluto Press, 2000.

Belden C.: *Landscapes of the Sacred: Geography and Narrative in American Spirituality*, New York: Paulist Press, 1988.

Berry, J.K. : *Beyond Mapping, Concepts, Algorithms and Issues in GIS*, Fort Collins, CO: GIS World Books, 1993.

Bolstad, P. : *GIS Fundamentals, A first text on Geographic Information Systems, Second Edition*, White Bear Lake, MN, Eider Press, 2005.

Burrough, P.A. and McDonnell, R.A. : *Principles of Geographical Information Systems*, Oxford University Press, Oxford, 1998.

Carrillo, J., Lung, Y., and van Tulder, R. : *Cars, Carriers of Regionalism*, Palgrave Macmillan, New York, 2004.

Cronon, William: *Changes in the Land: Indians, Colonists, and the Ecology of New England.* New York: Hill and Wang, 1983.

Davis, Mike: *Ecology of Fear: Los Angeles and the Imagination of Disaster.* New York: Vintage, 1999.

Dibbern, Jens : *The Sourcing of Application Software Services, Empirical Evidence of Cultural, Industry and Functional Differences,* Physica Verlag Heidelberg, 2004.

Elangovan K. : *GIS, Fundamentals, Applications and Population Geography*, New India Publishing Agency, New Delhi 2006.

Harvey, David: *Explanation in Geography.* London: Edward Arnold, 1969.

Harvey, Francis : *A Primer of GIS, Fundamental Geographic and Cartographic Concepts,* The Guilford Press, 2008.

Jackson, Peter: *Maps of Meaning: An Introduction to Cultural Geography.* London and Boston: Unwin Hyman, 1989.

James, T.: *The Best Poor Man's Country: A Geographical Study of Early Southeastern Pennsylvania.* Baltimore: Johns Hopkins Press, 1972.

King, Anthony D.: *Buildings and Society: Essays on the Social Development of the Built Enviroment.* London: Routledge & Kegan Paul, 1980.

Lewis, Earl. *In Their Own Interests: Race, Class, and Power in Twentieth-Century Norfolk, Virginia.* Berkeley: University of California Press, 1990.

Lewis, Peirce. *New Orleans: The Making of an Urban Landscape.* Cambridge, MA: Ballinger Publishing Co., 1976. Annotation

Michael P.: *The Making of the American Landscape.* Boston: Unwin Hyman, 1990.

Michael, H.: *City Form and Natural Process: Towards a New Urban Vernacular.* London: Croom Helm, 1984.

Neil, H.: *The Land of Contrasts: 1880-1901,* The American Culture 5. New York: G. Braziller, 1970.

Peter, G.: *The Slow Plague: A Geography of the AIDS Pandemic.* Oxford and Cambridge, MA: Blackwell Publishers, 1993.

Ray Allen: *The Far Western Frontier, 1830-1860.* The New American Nation Series. New York: Harper, 1956.

Richard SCASE : *Managing Creativity,* Open University Press, 2000.

Ritchie, J.: *Landscape and Material Life in Franklin County, Massachusetts, 1770-1860.* Knoxville: University of Tennessee Press, 1991.

Stephen, D.: *Fields of Vision: Landscape Imagery and National Identity in England and the United States.* Cambridge: Polity, 1993.

Victor, A. : *Geography of Art and Culture,* Contributions to Economic Analysis, Elsevier, 2004.

Watkins, Eric: *The Middle Eastern Environment,* London, The British Society for Middle Eastern Studie, 1995.

Wolf, Aaron T.: *Hydropolitics along the Jordan River, Scarce Water and its Impact on the Arab-Israeli Conflict,* NYC, United Nations University Press, 1995.

Index

M

N

O

P

R

S

T

U

V

W

❑❑❑